TRAMMEL'S TRACE

Gary Pinkerton

NUMBER FIVE
Red River Books, sponsored by
Texas A&M University–Texarkana
Doris Davis, General Editor

TRAMMEL'S TRACE

FROM THE NORTH

GARY L. PINKERTON

TEXAS A&M UNIVERSITY PRESS
COLLEGE STATION

Copyright © 2016 by Gary L. Pinkerton
All rights reserved
Second printing, 2018

This paper meets the requirements of ANSI/NISO Z39.48-1992 (Permanence of Paper).
Binding materials have been chosen for durability.
Manufactured in United States of America

LIBRARY OF CONGRESS CATALOGING-IN-PUBLICATION DATA
Names: Pinkerton, Gary L., 1954- author.
Title: Trammel's Trace : the first road to Texas from the north /
  Gary L. Pinkerton.
Other titles: Red River Valley books ; no. 5.
Description: First edition. | College Station : Texas A&M University Press,
  [2016] | Series: Red River Valley books ; number five | Includes
  bibliographical references and index.
Identifiers: LCCN 2016033459| ISBN 9781623494681 (cloth : alk. paper) |
  ISBN 9781623494698 (e-book) | ISBN 9781623497903 (paper)
Subjects: LCSH: Trammel's Trace (Tex. and Ark.) | Trammell, Nicholas,
  1780-1856. | Texas—History—19th century. | Texas—Colonization. |
  Arkansas—History—19th century.
Classification: LCC F392.T68 P56 2016 | DDC 976.4/05—dc23 LC record
  available at https://lccn.loc.gov/2016033459

CONTENTS

Preface
vii

Acknowledgments
xv

Chapter 1
From Somewhere to Nowhere
1

Chapter 2
Through the Wilds
5

Chapter 3
The Trammells of Kentucky and Tennessee
34

Chapter 4
1800–1812: Boundaries under Pressure
52

Chapter 5
1813–1819: Couriers of the Forest
75

Chapter 6
1820–1826: Gone to Texas and Back
99

Chapter 7
1826–1836: A Great Movement of Many Nations
134

Chapter 8
1836–1844: Another New Nation for Texas
166

Chapter 9
1845–1856: The Old Smuggler Retires
192

Chapter 10
1856–1880: The Patriarch Has Passed
214

Notes
223

Bibliography
259

Index
271

PREFACE

"Trammel's Trace, what is that?" I asked my father some ten years ago. Although I had grown up in East Texas, I did not remember ever hearing about Trammel's Trace. The families of both my parents were from the Mt. Enterprise area, and we still owned land near there. I don't remember how Trammel's Trace came up, but I clearly remember his answer.

"You know that old road rut across the pasture in front of the farm house? That was part of Trammel's Trace."

"You mean the one where we used to play hide and seek with the cousins, and build forts, and rake out trails in the leaves?" I asked, knowing full well which rut he was talking about.

"Yep."

"So what is Trammel's Trace?" I asked again. All he knew was that it was an old road into Texas and some guy used it to smuggle horses. So I did what anyone would do with their curiosities these days . . . I went to Google.

There wasn't much available, but the *Handbook of Texas Online* told me enough to set the hook on the bait my father dangled. It was indeed a *very* old road from the Red River to Nacogdoches that crossed right through our family land. It was over two hundred years old just since Anglo use, and who knows how long the trail had been used by Caddo and other tribes. I learned it was named for a horse smuggler named Nicholas Trammell.

With apologies to my seventh-grade Texas history teacher at Forest Park Junior High in Longview . . . I didn't know! Just think what a star student I could have been in Ladye Bird Taylor's history class if I had been able to bring pictures of a road used by smugglers, freebooters, and Texas heros. History classes had never managed to gain my full attention, but when I realized that Sam Houston, David Crockett, and Jim Bowie might have taken a rest break *right out there* in a rut across our pasture . . . well, that was pretty neat.

As if to reinforce what I had missed growing up, the Texas Historical Marker right in front of the cemetery where my grandparents are buried, which I had driven by hundreds of times, also said that Trammel's Trace ran along that very road right in front of the Shiloh Baptist Church.

And so this work began. Humbly. A curious yet apparently oblivious guy gets the opportunity to research a subject where so few facts had been gathered and the reward of being a topographic detective in the field. Not only were there facts to learn, but there were existing ruts to find all along the 180-mile length of the trace. It turns out that a few people knew a little about the trace, at least the myths, but no one knew very much. Now I get to share some of my research with the reader. If you are like me, you will learn things you didn't know about where you or your ancestors grew up and how they may have come to Texas and you will be intrigued by the connection to history that remains in the form of wagon ruts across a roadway or through a pasture all across East Texas.

Trammel's Trace, What Is That?

Trammel's Trace was the second major route into Spanish Texas from the United States and the first route from the northern boundaries along the Red River. In the early 1800s, Trammel's Trace was a smugglers' trail but later became a path for immigration to Texas. It was an historic corridor connecting travelers from Kentucky, Tennessee, Missouri, and Arkansas with the El Camino Real at Nacogdoches.

Trammel's Trace ran to Nacogdoches from two points on the Red River—Fulton, Arkansas and the early Pecan Point/Jonesboro settlements to the northwest of Clarksville, Texas. In Nacogdoches, it connected with the El Camino Real, also called the Old San Antonio Road, running east and west. Its history from the early 1800s through Texas' statehood is the history of migration, lawlessness, and conflict that defined that period. It is those stories about the land through which it passed and the people who traveled it that I hope to convey to the reader.

The road's namesake, a Tennessean named Nicholas Trammell (1780-1856), is the subject of much myth and legend. Although Spanish and Mexican authorities attempted to control trade in the region, smugglers found a way around the patrols. Nicholas Trammell was such a key part of much of the trade and early migration that a series of old trails linked together by his frequent travel were named after him in the early 1800s.

While the later stories of Trammell as a murderous outlaw are not supported by evidence, his life of smuggling and racing horses, operating taverns, gaming operations, and other opportunistic business dealings placed him on the fringes of frontier culture.

After Texas became a province of Mexico in 1821, the reasons to use Trammel's Trace shifted. Immigration, not trade, became the primary force of change. Trammell recognized that opportunity and moved from northeastern Arkansas to a trading post on the Red River in 1822 and later to a site near Nacogdoches. His original intent was to settle in the new colony started by Stephen F. Austin, but as a result of Trammell's poor reputation, Austin rejected his entry into the colony. He retreated to the Nacogdoches District, where he found a new opportunity at the Trinity River ferry crossing of the El Camino Real, a key route for smuggling horses and trade goods.

Trammell's favorable post placed him in the middle of disagreements over land between the old settlers and the new entrants hungry for control of the vast lands in Texas. In 1826 his good fortune ended when he was chased out of Texas as a consequence of land disputes leading up to the Fredonian Rebellion, a preview of revolutionary battles to come.

Returning to Arkansas, his business vision changed yet again. Trammell and his sons placed themselves along roads crossing the full width of southern Arkansas, from the Mississippi River to the Red River and farther west into the settlements around Pecan Point in northern Texas, with a series of taverns and gambling houses on the doorsteps of the turmoil in Texas. Rather than transporting trade goods and horses across hundreds of miles of unguarded territory, the Trammells simply waited for business to pass by their doorway.

Nicholas Trammell was an old man by the time Texas statehood arrived in 1845, and his sons had their own business plans. One by one, the Trammells returned to Texas, this time settling near Gonzales and LaGrange. Old Nick was the last one of them to make the trip down Trammel's Trace one last time. When he died in 1856, an era died with him. Nicholas Trammell was witness to dramatic changes in Texas, taking business advantage every time a new nation controlled that expansive territory. More opportunist than outlaw, Nicholas Trammell made the most of the disorder. Both Trammell's Trace and its namesake made their own paths through Texas history.

The focus of this work is a parallel history of Nicholas Trammell, the road named for him, and the role of that road in smuggling and immigration across northeastern Texas and southwestern Arkansas.

Thousands of years of human history preceded the emergence of Trammel's Trace in the piney woods of northeastern Texas and southwestern Arkansas. The Indian trails that came to be known as Trammel's Trace were key transportation corridors from the earliest habitation of man in the region. The opening chapter propels the reader through history toward the beginnings of Trammel's Trace and illuminates the footprints of the past leading to a journey down the old trail.

A common question from anyone interested in learning more about Trammel's Trace is "where did it go?" The second chapter addresses that question and creates a living image of the path of the trace through Arkansas and Texas. It describes what travel along the trace might have been like during the early 1830s. Not only does this chapter describe the path of the old road, but a day-by-day account of the conditions and terrain invites the reader to participate in the 180-mile journey.

Following the first two introductory chapters, the reader is presented with a history of Trammel's Trace and historical events in early Texas through seven distinct periods. Each period of time is a unique perspective on the life of Nicholas Trammell and the role Trammel's Trace played in the unsettled conditions of early East Texas.

The Trammell family always seemed to be on the leading edge of new settlements, and understanding the early life of the trail's namesake provides context for what is to come. After settling along the Kentucky-Tennessee border in the late eighteenth century, the Trammells and their kin made their way to what is now northeastern Arkansas, along with groups of other families. Horse racing and trading made their presence along the frontier fringes welcomed by the settlers and profitable for the Trammell family.

In the first decade of the nineteenth century, a cascade of political, cultural, and natural events led to the emergence of Trammel's Trace in response to the times. Smugglers and settlers could not resist crossing whatever boundaries existed. Trammel's Trace was a product of that insatiable taste for trade and settlement. Pressures focused on the old settlement of Nacogdoches from the north, east, and west set the stage for the use of Trammel's Trace for the movement of people and goods.

Smuggling was the prevailing economic engine in the early 1800s. Trammel's Trace was initially a smugglers' trail, a path to move contraband and stolen horses back and forth across the border into the United States. During this time period, conditions were favorable for illicit commerce and Trammell took advantage. He was able to roam freely across almost two hundred miles of territory with no settlements or soldiers, skirting the edges of the frontier and the reach of the law.

After Spanish Texas became Mexican Texas in 1821, waves of migration down Trammel's Trace led Nicholas to seek more traditional means for making a living. When Stephen F. Austin began to colonize Texas, Trammell, like hundreds of others, sought land in the colonies of Central Texas. Unlike many others, he was refused entry due to his reputation and was ultimately chased out of Texas at the point of a rifle.

Tensions increased and conflicts developed with the influx of Anglo settlers in a country still in dispute among Indians and Spaniards. The Fredonian Rebellion in Nacogdoches in 1826 was one of many minor conflicts that were precursors to the Texas Revolution.

After being run out of Nacogdoches in 1826, Nicholas Trammell returned to Arkansas to operate a tavern and engage in gambling and horse racing. Age began to slow down his questionable activities, but not the stories about his secretive, mysterious character. Trammel's Trace evolved from a smugglers' trail to a postal route; from the only road into Texas from the north into a county line.

At the end of his life, Nicholas Trammell and his family braved a move back to Texas around 1845, but they settled far to the west of Nacogdoches in La Grange. Even after Trammell died in Gonzales County, Texas, in 1856, his name lived on in court battles among the family over his estate and in myths and legends about his life.

Trammel's Trace Today

Although the history of Trammel's Trace as an active route took place two hundred years ago, there is a present-day story to Trammel's Trace as well. Parts of Trammel's Trace still exist across northeast Texas, and preserving them should be the goal of every landowner who recognizes their own piece of Texas history.

When the Republic of Texas surveyed the original headright land grants, the plat maps and field notes indicated the crossing of roads and trails at each boundary. Trammel's Trace was so marked and named on many of the surveys between the Red River and Nacogdoches. By converting the existing unit of measure—an old Spanish accounting of distance—it is possible to determine how far from the corner of a survey boundary the old Trace crossed. Using today's tools for GIS (geographic information systems) and mapping software (such as Terrain Navigator Pro), I have been able to locate precisely the ground where much of the old road crossed. Satellite images in Google Earth often confirm my findings through surface features that mirror the route of the trail. Ruts across a pasture too deep to mow leave a tree-lined marker to the path beneath them, visible using satellite imagery. Soil maps and topography also provide hints about how the trace followed a fault line, an outcropping of firmer ground, or curved around an elevation rather than crossing over the top of a hill.

On many occasions, after doing all the research and mapping work at home, I've been rewarded in the field by walking to that specific spot in the woods or along a roadside and finding the old route of Trammel's Trace beneath my feet. It was quite an experience once I learned more about the stories of what took place along that route. The ferry crossings and fords at Ramsdale's Ferry on the Sabine and Epperson's Ferry on the Sulphur gave me chills when I visited for the first time. I could imagine all the people who crossed those rivers after miles of hard travel. No doubt there were many camps made along the banks to enjoy a respite or wait for floodwaters to recede.

Although this book will find interest among current residents of East Texas and those from other states whose ancestors made the trek to Texas, it is really the landowners for whom I hope this work strikes a particular interest. There are remains of Trammel's Trace all across Bowie, Cass, Marion, Harrison, Rusk/Panola, and Nacogdoches Counties in Texas and Little River county in Arkansas. Not every rut will be Trammel's Trace, but if the surveys tell you where to look and when you go there you find a rut, you can be pretty certain that Sam Houston might have taken a break on your land as well. My fellow "rut nuts" and I will be happy to help you. I have met many wonderful people in the process of producing this work, and there are many more to meet.

Trammel's Trace ruts in Cass County. Photo by author.

ACKNOWLEDGMENTS

In the years of research that have gone into this current work, many people have been helpful in the project. In this space I can only call attention to a few who have been there for key milestones. Gail Martin at the Southwest Arkansas Regional Archive in Washington, Arkansas, unfortunately did not live long enough to see this work emerge. My early contact with her led to the discovery of the Trammel's Trace research of James and Mary Dawson from the 1940s. James Dawson was a civil engineer and an expert surveyor and draftsman who did extensive research on the location of the trace. Some of the maps in this work are based on his earlier effort. Gail was incredibly nice and helpful and is missed by everyone who came to know her.

Sue Lazara, Bob Vernon, and Desiree Bryant were all present for the first field trip based on my research. Their field talents included helping push my truck uphill over red, rain-slicked Cass County mud. Bob is a Texas Archaeological Steward, a flintknapper, and generally one of the smartest and most talented people I have ever known. Bob's computer skills in illustrating some of our fieldwork has been exceptional. If Bob's energy determines his lifespan, he will live to be about 184.

One of the many archivists and historians and county clerks I have worked with offered this quote by Steven C. Levi: "History . . . is not a re-creation of the past. It's an assessment of the past based on documents provided by people in archives and museums who will answer your letters."

The direction and outright hand-holding that each of them has offered has been critical to this project. Holding early county records in my hand, getting close-up looks at early Texas maps so wonderfully preserved and digitized by the Texas General Land Office, and gleaning the tiniest bits of information after gaining their interest and support have been wonderful benefits of this work.

Finally, I have to acknowledge the support of the two most important women in my life, my mother, Joan Strong Pinkerton, and my wife, Mickey Arnold Hammond. Their gentle, and less than gentle, nudges and their tag-team approach to ganging up on me has resulted in a work we are all happy to see emerge.

Although the book is done, it will never be finished; especially the work in the field, educating residents of the counties where Trammel's Trace remains, meeting landowners, and telling the story of these early immigrants. Modern threats to what may appear to most as nothing more than a depression in the earth continue. Strip mining of the same lignite coal deposit that made the Sabine River crossing a solid base for horses and wagons has taken place for miles on both sides of the Sabine. Most recently, the entire area at Ramsdale's Ferry was mined and a road built large enough for the mega-cranes to cross.

Although that kind of destruction is extreme and dramatic, it is no more certain than a landowner filling in ruts with trash or plowing them level. My hope is that this book will inspire a renewed interest in Trammel's Trace and a passionate effort on the part of landowners to preserve their piece of history.

Strip mining along Rusk Co. line obliderates reamains of Trammel's Trace. Photo by author, 2010.

TRAMMEL'S TRACE

1

From Somewhere to Nowhere

When I plunge into chasms, sound the deeps,
Climb the plateau, mount the dizzy steeps;
Or when above the mountains, near the skys,
I spread new worlds before man's dazzled eyes,
For all the gifts on you I have bestowed,
Be my good doctor—save, oh save the road.

—HELEN NORVELL[1]

For all the places people now travel, someone was the first to be there, the first to observe the natural beauty, the first to gaze across a grand vista. Someone was the first to stand on a bluff at the edge of a broad river and wonder from where such a mighty force could come, where it might go, or how to get across. Others found places of sacred beauty and came to stand in awe. More practically, hunters found plentiful game or trade in a new direction and were rewarded for their travels. As others followed, their footsteps wore paths to springs or honey trees, to shelter and safety, or to sacred places where the Great Spirit was present.

Native people first ventured to these places in northeastern Texas and southwestern Arkansas around 11,000 BC. Small groups on hunting expeditions tracked bear, deer, and buffalo, following animal signs and hoof paths for hundreds of miles. They built simple shelters, easily abandoned when it was time to follow the game once again. They moved across vast prairies and through deep woods in search of what the land might provide. They left behind little more than well-worn paths.

Over the centuries, greater numbers of inhabitants followed trails once trod by bare feet. Villages appeared at sites offering security and sustenance. The people in these villages built more lasting homes of cane, grass, and mud. Peaceful Native Americans from the north, the Caddo people, settled East Texas forests and developed a society around their agrarian lifestyle. Their burial mounds have remained for hundreds of years in testimony to their culture and ritual.[2] They cut clearings from the forests with stone axes. Plots for squash, sunflowers, maize, and beans appeared along the edges of the dense forests. People who traveled the trails planted peaches and grapes to serve as nourishment during their travels and markers along the way. Pathways worn by regular use became trails to neighboring villages and springs or to distant trading partners. Trading cultures evolved, and long-distance travel for hunting and trade became the primary reason for continued use of these ancient pathways.

Trade activity widened the woodland trails of East Texas long before Europeans ever came to the New World. It was much later when kings seeking riches of gold and an expansion of their empires sanctioned expeditions to the newly discovered continent far across the Atlantic Ocean. Those who survived the long voyage and difficult months in the wilds encountered the trails worn by the people who had lived there for centuries. When they arrived in what is now eastern Texas, Spanish explorers found existing paths wide enough to allow the passage of Spanish horsemen, soldiers, and livestock.[3] Spanish explorer Álvar Núñez Cabeza de Vaca made his way to the region between 1528 and 1533. Hernando de Soto and a contingent of six hundred men landed on the west coast of Florida in 1539 and explored the Gulf Coast for four years. After de Soto died, in 1542 an expedition of survivors led by Luis de Moscoso Alvarado crossed the Red River and traveled trails along a geologic uplift in eastern Texas, searching for an overland route to New Spain. In 1685, French explorer Robert La Salle encountered native people, likely Caddo, near Nacogdoches. Spain established the first two European missions west of the Neches River in 1690, followed by four missions among the Caddo on the upper Attoyac Bayou in East Texas and two at Nacogdoches before 1716. The hooves of Spanish ponies traced the trails of buffalo and deer, and the boots of soldiers and missionaries followed footpaths left years before.

The territory that became Louisiana and Texas was at the center of a tense imbalance between France, Spain, and the Indian nations in the mid-eighteenth century. Athanase de Mézières, a French soldier stationed in

Natchitoches, Louisiana, was one of the few men who maintained good relations with each of those nations. The Spanish government recruited de Mézières to strengthen relations with the Indian tribes to the north and west. He crossed the Sabine, Angelina, and Neches Rivers in East Texas toward San Antonio de Bexar in 1772, accompanied by seventy chiefs and tribal leaders on an assignment to help resolve Indian troubles.[4] He and his noble chiefs found their way using the same pathways and trails that Spanish traders and explorers used 150 years earlier. De Mézières, in a letter dated May 20, 1770, observed the character and behavior of traders and settlers in the region and provided his assessment of the difficulties: the region "being the asylum of the most wicked persons, without doubt, in all the Indies. . . in short they have no other rule than their own caprice, and the respect they pay the boldest and most daring, who control them."[5] Men like those who lived in what is now southwestern Arkansas and northeastern Texas came in search of game, to initiate trade and commerce, to occupy and claim land, or simply to rob, cheat, and defraud those who pursued those ends legitimately.

Many distinct forces at the dawn of the nineteenth century led to increasing traffic over the early trails. In the years preceding the Louisiana Purchase in 1803, a different kind of traveler began to use the footpaths created by native people centuries before. Men on horseback from the United States were unable to resist the lure of a seemingly endless country. Immigrants moving west wore their own hoof prints and wagon ruts into barely discernable paths. The sounds of a trader's wares clanging and creaking on the back of a mule or the wild cries of mustangs trying to escape capture were heard in places where few white men had traveled. In the silence of these wilds, the only sound one heard was the "resonance of their living being."[6] Now there was the clamor of change that could be widely heard.

The El Camino Real de los Tejas, Spanish Texas' only major road of the time, ran east and west through Natchitoches, Louisiana, toward Nacogdoches and San Antonio in Spanish Texas and on to New Spain.[7] Also known as the King's Highway or Old San Antonio Road, it carried traders and old settlers between the United States and Spanish territory. Although the El Camino Real provided access from the United States to the east, there was no north-south road into Spanish Texas in the early 1800s. Only an Indian trace from the Great Bend of the Red River, in what is now southwestern Arkansas, toward the El Camino Real at Nacogdoches led to Texas from the north. Its route was scattered with the castoff arms and utensils of

Spanish soldiers from the past and bounded by ancient burial mounds. Around 1813, the forces of history began to define this network of pathways as the first road into Texas from the United States to the north.

Over time, the road to Spanish Texas from the north gained favor with people from Arkansas, Missouri, Kentucky, and Tennessee who sought land west of the Louisiana Purchase. A series of trails across Arkansas, later called the Old Military Road or Southwest Trail, led all the way from St. Louis on the Mississippi River across Arkansas to the Great Bend of the Red River, where they connected with the trail into Texas northeast of what is now Texarkana.

With access becoming easier, the borderlands north and east of Nacogdoches became a haven for criminals, outlaws, and suspect men of all kinds. Stealing horses and trading in contraband were part of the commerce of the day. Boundary disputes, fraudulent land claims, international conflict, and a culture of lawlessness dominated the region. At the disputed boundary between the United States and Spain along the Sabine River east of Nacogdoches, military leaders in a tenuous standoff in 1806 negotiated an agreement to set aside a "neutral ground" between the nations. Neither country enforced order in the resulting borderlands, leaving behind a safe haven for criminal activity. The trail south into Spanish Texas from the Arkansas Territory led to that neutral ground. It became a smuggler's back alley in a no-man's land at the edge of two nations.

Early trails and roads were there for a reason. The people who originated them and those who traveled them later did so with a purpose. The more efficient or direct routes were more widely traveled. Trails became pathways of choice only because of the practicality they offered, communicated by word of mouth. Improvements were made by settlers and slaves or by opportunists and criminals. With no regular traffic to keep them clear, unused roads were quickly overgrown and disappeared back into the undergrowth. Their rutted paths remained the only evidence of their existence.

Slowly, over centuries at first, then accelerated by migration and commerce in the new frontier, the trail into Texas from the north became a road that directed thousands of people to their destinies. Some of Texas' most famous heroes, and its most infamous smugglers and outlaws, traveled this early pathway. In the early 1800s, that roadway from the north became known as Trammel's Trace.

2

Through the Wilds

*There had never been a road prepared
that a wheeled conveyance could pass
from one section to the other.*

—R. L. JONES[1]

The word *trace* is used to describe a path worn through the wilds by the passage of men or animals. A trace does not qualify as a trail. It is something less than that, something not quite as well defined or marked. The root word, *tracer*, has origins in both French and Spanish, much like the mix of cultures prevalent in the borderlands Trammel's Trace crossed in the early 1800s.

The trails that later became known as Trammel's Trace were adopted for regular use after 1813 by men moving horses and trade goods from the Red River prairies through Nacogdoches and across the Sabine River into Louisiana. Trammel's Trace from the north and the El Camino Real de los Tejas from the east were the earliest two routes into the Nacogdoches district of Spanish Texas from the United States.

Smugglers needed a road beyond the patrols of the authorities, and Trammel's Trace was ideally situated for that purpose. The removal of mustangs from Texas and the import of trade goods from the United States were both declared illegal by the Spanish crown. Vast herds of wild horses along the Red River and willing buyers in the United States provided ideal conditions for smuggling. Spanish authorities required Nacogdoches residents to trade for supplies, guns, and hard goods far to the west in San

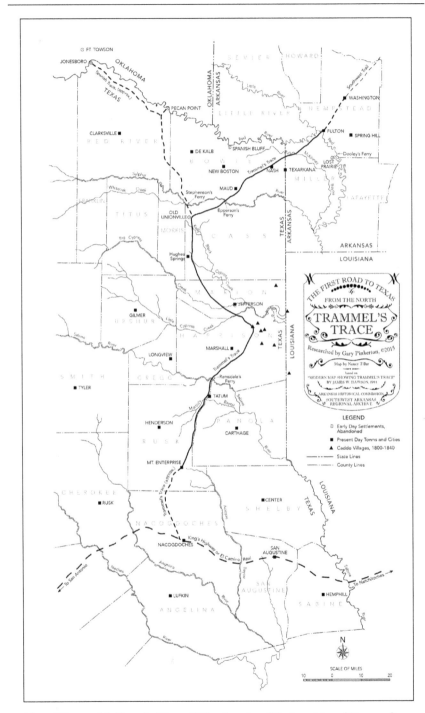

Map of Trammels's Trace by Nancy Tiller, based on "Modern Map of Trammel's Trace" by James Dawson. Courtesy of Arkansas Historical Commission.

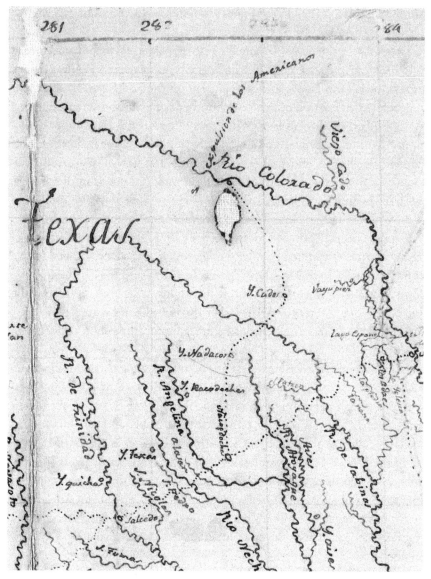

1807 map of Texas shows general route of part of trail that became Trammel's Trace from Red River to Nacogdoches. By Jose Maria Puelles, Texas Map Collection, The Dolph Briscoe Center for American History, The University of Texas at Austin.

Antonio, although the United States was a much closer and more lucrative journey to the east.

The earliest evidence of Trammel's Trace appeared on maps from the early 1800s. A map drawn in 1807 by a Spanish priest named Puelles showed a road north from Nacogdoches to a point on the Red River. In 1803, not long before the creation of this map, hundreds of Spanish soldiers marched to the Red River from Nacogdoches to intercept a US expedition exploring the boundary between Spain and the United States. Following trails through the forests and prairies, the soldiers left behind a scarred landscape clearly visible to others. Men who captured mustangs in the prairies south of the Red River later used that soldiers' trail. Settlers also used parts of the route to migrate into Texas many years later.[2]

Spanish Texas became Mexican Texas in 1821, and the liberalization of colonization laws resulted in Anglos migrating in growing numbers. People from Tennessee, Kentucky, the Missouri Territory, and the Carolinas came down Trammel's Trace from both of its origins at separate points on the Red River—one at Fulton, Arkansas, at the Great Bend of the Red River, and the other from settlements at Pecan Point and Jonesborough, farther west up the Red River along what would become the border between Texas and Oklahoma.[3] Increasing numbers of immigrants in the 1820s overwhelmed the weak and overextended Mexican government, adding pressure to establish communities and social norms more reflective of the new America than the old Spanish province. Conflict between the immigrants and the government in their new country was inevitable, inflamed by a belief in the United States' manifest right to spread its control across the continent. Nacogdoches was at the crossroads of that change, both literally and figuratively, the only established settlement in Texas east of San Antonio de Bexar. It was no accident that Trammel's Trace terminated in Nacogdoches and became a primary path to events that led to the formation of a new republic.

In anticipation of business ventures in Texas, in 1819 Stephen F. Austin (often called "the father of Texas") and his father, Moses Austin, acquired land at Long Prairie in what is now southwestern Arkansas. Their advertisement for lots in the *Arkansas Gazette* described the trail as the easiest point of entry into Texas. That entry point was Trammel's Trace.[4] The Mexican government permitted Austin to bring Anglo colonists into Central Texas in 1821. As a requirement of his contract as *empresario*, Austin created maps of Texas in 1822 and 1829. Austin's 1829 map showed a clear path that

Mapa Original de Tejas/por el ciudano Estavan F. Austin; Draftsman, Stephen F. Austin 1829; Texas General Land Office (GLO) Map #2105.

matched the route of Trammel's Trace from southwestern Arkansas and included a branch toward Pecan Point and Jonesborough.[5] His map showed the northeastern branch of Trammel's Trace heading to Long Prairie, just south of Fulton. Austin had direct knowledge of the road from there into Texas.

Although the earliest maps did not name the trails across northeastern Texas, the track they sketched matches later maps, which attached the name Trammel's Trace. The earliest mention of Trammel's Trace by name was in a letter dated June 1821, when an early Red River settler referred to the "old Trammel trace."[6]

Subsequently, Trammel's Trace was firmly entrenched in the cartographic history of the developing region. When the Texas Republic began making grants of land in 1838, surveyors noted the crossing of Trammel's Trace through many of the original Texas headright grants in seven counties. Commissioners forming the boundaries of Rusk County, Texas, in 1843 designated Trammel's Trace as two-thirds of the line between Rusk and Panola counties. Even into the mid-1860s, Civil War cartographers identified Trammel's Trace amidst a growing network of roads crisscrossing the region.[7] In places where later roads followed on or near the original path of Trammel's Trace, evidence of the old road remains today in the form of overgrown ruts through forests or across cleared pastureland.

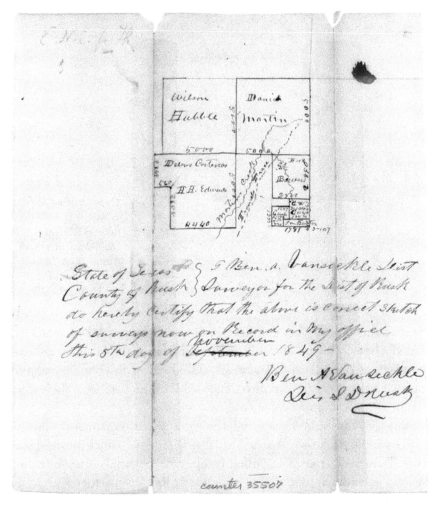

Daniel Martin survey, 1849. Texas General Land Office. This document, "Rusk County Sketch File 10, Sketch on Martin Creek, northeast of Henderson, GLO scrap file 35507" is from the B. A. Vansickle headright survey made on November 5, 1849.

A Journey down Trammel's Trace

Travelers making their way southward along Trammel's Trace for the first, and likely the only, time had no maps or guides. A narrow trail through virgin forest or across a muddy river bottom could not be detailed on any map of that era. The best sources for guidance were descriptions of the route from others who traveled it. Depictions of Trammel's Trace emerged from the letters and diaries of early settlers and travelers, official reports by envoys of Spain, Mexico, and the United States, and the daily logs of soldiers patrolling the edges of the frontier.[8]

Roads led from one distant point to another and were sometimes marked with blazes or by paintings on trees. The uninitiated could never tell if a side trail might lead to a spring, bubbling clear water out of the ground, or to the remains of a passerby robbed and killed for his horse. Tired, hungry, and unsure, travelers saw every fork as a decision point where the wrong choice might send lost souls into peril toward some hostility or into the hands of land pirates. Uncertainty was no friend of travelers, who were always on guard against dangers, human and otherwise.

To get to the northern border between the United States and Mexico in the early 1820s, immigrants first had to cross the Arkansas Territory. The road across Arkansas cut diagonally from northeast to southwest, following a geologic boundary between the hills to the north and west and the

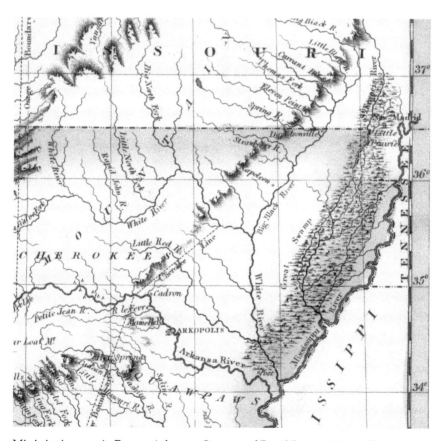

Mississippi swamp in Eastern Arkansas. Courtesy of David Rumsey Map Collection.

plains to the southeast.⁹ Although the United States military had improved the trail, increasingly frequent use left it rutted and rough and forced wagons and horses around the remains of stumps eight inches or more above ground. In one section of the fearsome Mississippi Swamp in eastern Arkansas, only about five miles of high ground was found in ninety miles of river bottom.[10]

No matter what difficulties travelers encountered crossing Arkansas, their journey from the Red River down Trammel's Trace was unlike any other leg of the trip. The travel yet to come marked the most uncertain part of the journey. Stepping onto Trammel's Trace meant stepping onto a path that crossed through unsettled lands in a foreign country. At the pace of a loaded wagon, the trip from Fulton to Nacogdoches took roughly two and a half weeks. Two and a half weeks of forests, two and a half weeks of worry over attack by Indians, two and a half weeks when not a soul might be seen.

When immigration toward Texas began to increase after 1822, settlers headed to Texas in wagons loaded with all the possessions they could carry entered the region with little understanding of what to expect when they arrived. The day-by-day account that follows is based on both the geography of the route of Trammel's Trace and accounts from the period about means of travel. With a loaded wagon, fifteen miles a day was making rapid headway. That measure, as well as the geography, is used to provide an account that identifies not only the physical difficulties presented by the terrain but also the feelings travelers of that time were likely to experience.[11]

Day 1: Leaving the United States

The trip down Trammel's Trace began with fear and anticipation—a fear of the unknown and anticipation of the opportunity for hundreds of acres of land and a fresh start. The transition to the uncertain eastern boundary of Mexican Texas began fourteen miles northeast of the Red River at the crossroads in Washington, Arkansas, where travelers learned there were two primary routes into Texas. Trammel's Trace crossed the Red River directly into what is now northeast Texas by way of Fulton, Arkansas. Earlier travelers continued farther west up the Red River to use the Trammel's Trace branch from Jonesboro and Pecan Point. That road was only recently cleared for use by wagons.[12] The second major route to Texas followed the east side of the Red River to Natchitoches, Louisiana. The advantage of that route was that it remained in the territory of the United States. By comparison, Trammel's Trace was reported to have better access to water and to be a faster journey.[13]

The United States' interest in turning the old trail across Arkansas into a military road clearly stopped at Washington. Travelers could sense that they were venturing beyond the reach of the United States on this final stretch toward Trammel's Trace at the Red River. They had to be self-reliant and vigilant against the hazards of their continuing journey southwest into Texas. Even if the road was widened from regular use, wagons could soon break down from the rough terrain. In a letter to his family back in North Carolina, Jeremy Hilliard wrote after his trip down Trammel's Trace of how his newly built wagons began to fall apart only days into his trip: "The iron works on my waggon have bothered me more than everything else bursting and buckling to pieces. I fear it will turn out to be a dear wagon to me. I never was worse treated in a piece of work. Amelia says tell Mr. Hill when he moves don't let Mr. Stalings have anything to do with his wagons."[14]

Traders and immigrants also used cowhide packsaddles to carry their belongings well before the Trace was cleared for wagons. As long as the space between a horse's legs, the pack was strapped across its girth. These Mexican-style packs, called *sillajes*, were big enough to carry one of the most prized possessions on the frontier—a featherbed.[15]

Fulton, at the Red River, is fourteen miles southwest of Washington, Arkansas, near enough to make the trip in one day of travel. The trail from Washington to Fulton came down the southwest face of the foothills. Just above a bluff along the river, the single trail split into a series of small traces. Each trace led to separate campsites and river crossings, where various tribes and traders dispersed to avoid encroaching on each other.[16] A sunset camp on the high banks at the Red River at Fulton was the reward for those who pulled out of Washington at first light. At the end of a long day, travelers could stand on a bluff overlooking their next obstacle, the crossing of the Red River. They were about to traverse 180 miles of uninhabited, foreign country before emerging in Nacogdoches. A good night's sleep in the relative safety of the United States was a comfort that would soon end.

Day 2: Crossing the Red

Riverbanks are designated as right or left from the perspective of a boatman facing downstream. Fulton, the oldest settlement in southwestern Arkansas, was founded on the left bank of the Red River where the water slowly eroded the hillside in a grand turn from an easterly course to almost due south—the Great Bend. River access, a road to the interior of the Arkansas Territory, and fertile lands were key assets for Fulton. Developers believed Fulton could become a prosperous place for commercial ventures. In spite of the possibilities, few stopped to settle there, particularly with all the land farther south in Texas seemingly there for the taking.

When the sun rose behind the last campsite in the United States at Fulton, it illuminated a sought-after destination on the other side of the river. Sounds of the morning coming to life eased the minds of travelers otherwise occupied by the difficulties of the journey they were about to undertake. There were no inhabitants along the way who might offer safe respite for a night. The only grain their horses would have was what they could carry. The likelihood of danger and adversity was common knowledge.

Crossing the Red River could be simple or treacherous. Shoals and shallow spots made any ford much easier. Natural crossings built up where smaller streams entered the flow of a larger river. A mile or so upstream from Fulton, where the Little River merged its waters with the Red, sand

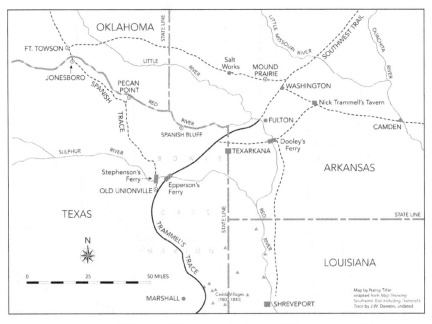

Map of northern half of Trammel's Trace. Map by Nancy Tiller, based on "Modern Map of Trammel's Trace" by James Dawson. Permission of Arkansas Historical Commission.

bars and islands of debris created a ford that for much of the year was the easiest way across.[17] Animals found the most favorable fords. Their historic paths were the best guides to suitable crossings. Just downstream of the bluff at Fulton, signs of horse's hooves angling down the bank to the water's edge indicated a manageable crossing. Even if the water level appeared passable, thorough reconnoiters were needed before making a choice about where to cross the ever-changing river. If the water was running swiftly from heavy rains, the boiling waters of the Red were tinted with the color of the clay soil through which they flowed. A darkened swirl two hundred yards across convinced the sensible to wait until water levels receded.

Any traveler arriving at the Great Bend from the United States had already made several river crossings along the way. At the edge of any substantial river, a man on horseback might swim or ride across to test the waters. Horses faced with a rapid, muddy flow hesitated before stepping off into water that could be up to their withers. There was no turning back at that point, even if it was deeper than it appeared; there was only time for a nervous slap of the reins to move forward.

A passage without wagons made the crossing easier. Packs were piled atop the tallest horse to prevent them from getting wet, or they were towed across on a small log raft built only for that purpose. A raft built by an earlier party could be reused, but only after it was retrieved from the other side of the river. If the water level was low, a wagon might still be pulled across by the horses. A passage with a wagon increased the danger and difficulty and put more belongings at risk. There was always a danger that the current might push against the sideboards or sweep over the top, swamping the wagon and pulling the horses under the current. Carts and mules had been lost in similar circumstances, even when the flow did not seem that strong. Eddies and currents caused the sandy bottom to shift underneath the horses' feet, and the force of the flow could overcome a horse or topple a wagon.[18] Supplies were unloaded from the wagon and moved across on small rafts. Only the inexperienced or foolhardy made a treacherous crossing with a wagon loaded with their only possessions. To get flour wet this early in the trip down the emptiness of Trammel's Trace would be unfortunate. To get gunpowder wet would be a disaster.

If the water was deeper, wagon boxes were raised by placing logs between the bottom and the running gear to gain a foot or so of height. If the water was too deep for the wagon to be pulled across at all, the wheels were removed and trees felled to build a log raft to carry it across. Traveling in a group made many things easier. With enough men, wagons that had been tarred and sealed, helping them to float, could be pulled across with a rope made from hide.

Safe arrival on the other side did not mean the dangers were over. What looked like a wide beach on the other side of the Red could in fact be a deep bog that shook with every step. After piercing the thin, drying crust, horses sank deeper into the muck, struggling to keep their footing.[19] The muck in the bottomland on the other side sucked the energy from the horse's legs and clung to wagon wheels like tar. The driver, muddy to his thighs, coaxed and pulled and hollered the horses across with the reins in one hand and a piece of switch cane in the other. Animals, property, and people were soaked and covered in dried dirt after several trips back to recover supplies from the other side of the river. A crossing like this could take hours for a large group to travel only two hundred yards, consuming the better part of a day and making little forward progress.

Not far from the riverbank was a well-used camp for weary travelers, a minor reward for the crossing. Signs of cook fires a few days prior left their mark under the branches of an old weathered oak. Past the campsite, the path of Trammel's Trace quickly left the relative openness of the riverbank and entered a canebrake, which covered miles of the route. The sky above the trail to Texas was about to disappear.

Day 3: Breaking the Cane

Cartographers who mapped the area around Fulton during the Civil War labeled Trammel's Trace on the Texas side of the Red River with a simple one-word testament describing conditions on the trail—"impracticable."[20] After the challenges of the trail across Arkansas, it was hard to imagine the road getting worse, but that was likely. From all accounts, Trammel's Trace fit the definition of a trace as something less than a trail. The road leaving the Great Bend of the Red River simply felt different from the roads across the Arkansas Territory. A road's feel was about how closed in it seemed, how jittery it made the horses, and what kind of improvements were present, if any. The road across Arkansas was rough but still showed signs of a modest effort at improvement by the military and by settlers anxious to bring trade past their door. Trammel's Trace looked more like it had only just been beaten through the woods. The entire route showed signs that it had not been long since it was simply a trail for smugglers. Those signs were found in the thick canebrakes on the Texas side of the Red River.

Through the Wilds 17

Illustration of a group of people on horseback riding through tall cane from "Visit to Texas," 1834, road through a cane brake, di_06621, Briscoe Center for American History, The University of Texas at Austin.

Cane twenty to thirty feet tall and an inch or more in diameter grew so thick that a man on foot had to weave his way like a tick crawling through the hairs on the back of an old hound. Seven miles of canebrake stood in the way, a canopy so thick that the sun could not penetrate for four months of the year. When cane was flattened or cut for a passage, the tall, leafy tops on the remaining sides curved inward toward the center and formed a darkened archway.[21] This tunnel of cane was beaten down only by the regularity of travel.[22] Horses did not like the closed-in feeling in the darkened tunnel of cane. They flinched every time dry reeds popped and snapped, sounding like a small pistol being discharged under their hooves.

In other sections, hardwoods that survived the regular inundations in the broad floodplain of the Red River allowed tall shoots of cane to grow underneath. Clearing the canebrakes, passing Lake Comfort, and crossing McKinney Bayou, Trammel's Trace slowly angled southwesterly toward a long bluff overlooking the rich river bottom to the north. The slight elevation along the edge of the floodplain brought a refreshing breeze and a glimpse of the land spoken of so highly by others. The border between the United States and Mexico was still uncertain in this part of the country, but at this point travelers understood they were entering a foreign land as the sun set on their third day down Trammel's Trace.

Day 4: The Sulphur Prairies

Meager supplies of coffee were carefully hoarded, but what better occasion to enjoy coffee than waking to the first dawn in a new country? For the first time, the travelers beheld prairies bordering the Sulphur Fork of the Red River in a country that had been described as beautiful and undulating, equal parts open grassland and wooded forests over low, rolling hills.[23]

The origins of Trammel's Trace as an Indian path became more apparent as it emerged from the muddy floodplains and canebrakes. Native American routes clung to the high ground to improve vantage points. Trees and brush grew close to the trail during the spring, sometimes limiting visibility to no more than twenty feet ahead. Trails were often winding, avoiding places where travel was difficult; the trail builders sought out terrain that allowed some seclusion.[24]

The westerly course of the Trace followed a ridge at the edge of the Red River bottom. Years later the same land became part of the old Sugar Hill Plantation, and now it is a combination of farmhouses and subdivisions northeast of Texarkana, Texas. The high ground gave relief from the floodplains below and still provides an expansive hilltop view. A few more miles west, the path of the Trace crossed what is now the Arkansas-Texas state line, about two miles north of Interstate 30 on State Line Avenue.[25] When the survey of the eastern boundary between the Republic of Texas and the United States was completed in 1841, members of the crew built an eight-foot-tall mound of dirt every mile along the line. Survey notes identified Trammel's Trace crossing between border mound markers 102 and 103, intersecting another road from the broad river bottoms to the east and the Red River south of Fulton.

Beyond the ridge along the floodplain, Trammel's Trace turned to the southwest, where legend says smugglers kept a hidden stockade for stolen horses. What is now Bowie County, Texas, was a mix of intermittent grassland and hardwood forest. Wild horses often moved down from the northern plains into these prairies. The sounds of nearby mustangs attracted packhorses to freedom. The gentlest wagon horse, once among a free-running herd, quickly acquired all of the intractable wildness of his

Through the Wilds

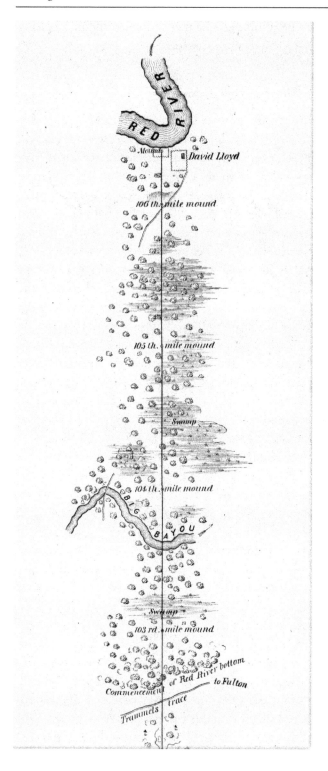

Trammel's Trace crossing of the Texas/US Boundary, 1841. When the boundary between the Republic of Texas and the United States was marked in 1840–41, eight-foot tall dirt mounds were constructed every mile along the line north of the Sabine at Logansport, La. Two roads crossed near there, Trammel's Trace and the road from Dooley's Ferry.

untamed companions. The result of such an encounter could be a bolting horse, nostrils flaring with the scent of freedom.[26]

Day 5: Creek Crossings

The ease of the trail through the prairies ended too quickly. Early in this day's journey, it was evident that a bluff on Nettles Creek was a transition point to a different kind of terrain. The trail headed downhill toward another water crossing—several of them, in fact. The East Fork, Nettles Creek, and Conn Creek had to be crossed in a distance of less than three miles.[27] Hardwood forests of oak, ash, and cedar, tangled with vines, lined the edges of the creeks. Small streams this close together resulted in continuous, thick undergrowth, making travel more difficult.

Some later travelers may have seen evidence of the trail's early use by smugglers in the form of puncheons laid across a creek bed to ease the crossing over soft ground. Puncheons are logs split lengthwise and recessed in the ground with the flat face upward. Puncheons could not be laid the whole way across this muddy bottom. Detours that formed around low, rutted spots in the trail took on the characteristics of side trails. Over time, some of those side trails replaced sections of the Trace and became the main path. Multiple and parallel tracks spread out across the terrain in places where wet or rutted trails made for rough going.[28] As the trail weaved its way from one high spot to the next, there were still sloughs in this bottomland that it could not avoid. Horses constantly sinking into the mud in this kind of terrain required wagon passengers to climb down and coax them through. Pulling on the horses and pushing the wagon wheels through the muddy bottoms resulted in slow progress. Even if a spot of dry ground to catch one's breath could be found, enduring the nasty biting insects made every stop painful and miserable. At least the clouds of black flies did not follow from the woods to the more open ground. When black flies swarmed the horses, they left a drop of blood at every spot they touched.[29]

The trail turned more to the south as it approached Caney Creek and Big Creek, separated by only about a mile and a half. It might take an entire day for travelers to slog through these creek bottoms. Relatively dry weather made travel easier than it was after an accumulation of rainfall. Wooded creek bottoms and bay galls (forested wetlands in a depression) did not dry out quickly, making for muddy going even without any recent rain. The land rose and leveled out again about a mile and a half past Big

Creek. Back on the edge of a small prairie, the trail emerged from the deep woods, where the clouds of mosquitoes relented for a change. The next oasis of trees at the edge of a little prairie was a good place to build a fire to cook a dinner of venison. Earlier on this day of creek crossings, it was easy to mistake this floodplain for the Sulphur Fork of the Red River, but these creeks were just a foreshadowing of a difficult crossing yet to come.

Day 6: Crossing the Sulphur River

Travel the next morning was easier, but only for a few miles. The dark mud from the sloughs dried and began to fall from the bottom of the wagon. The woods opened up into scattered prairies edged by mature hardwoods and pines. The ease did not continue, however, and the forest trail became thicker and more confined as the travelers entered yet another hardwood bottom. Traveling through similar country, one observer noted that when "the country began to descend a change soon took place in the aspect of nature, and of everything around us."[30]

Changes in the terrain were sometimes gradual and at other times sudden. Cross-country travelers learned to read the land as they made their long journey south. Watersheds felt different from one place to the next. The lay of the land, the plants and undergrowth, and the angle of the slopes told different stories. It was clear that this trail did not lead to just another creek crossing.

Tracks of horses and wagon wheels from earlier crossings were filled with a watery muck. The trail wound around haphazardly in the muddy sections, making way toward any higher patch of ground or spaces between trees wide enough to let packhorses or wagons through. Danger was camouflaged and hidden in the little oases of land. Alligators coated with dried mud hid easily, so numerous in places that a river traveler once shot at over forty in an hour.[31] One diarist chronicled the difficulties this way: "The unfortunate traveler has but little chance of escaping with life, if, from want of experience, he is foundered in the swampy cane-brakes. When the horse sinks and the rider leaves the saddle, the only thing he can do is to return back upon his track; but let him beware of these solitary small patches of briars, generally three or four yards in circumference, which are spread here and there on the edges of the cane-brakes, for there he will meet with deadly reptiles and snakes unknown in the prairies; such as the grey-ringed water moccasin, the brown viper, the black congo with red head and the copper head, all of whom congregate and it may be said make their nests in these little dry oases, and their bite is followed by instantaneous death."[32]

The approach to woodland rivers like the Sulphur was across boggy, miry, nasty ground. Horses tired quickly in this kind of terrain and required time to recover and rest. The left bank of the Sulphur River finally appeared and was the first river crossing encountered since the trip down Trammel's Trace started at the Red River. The trail led to a low bluff with a steep incline down to the river. The river was no more than fifty feet across, but the banks dropped off quickly, making the crossing more difficult. For two miles on the west side of the Sulphur River crossing, the floodplain was as flat as the bottom of an iron skillet, with virtually no change in elevation or the unrelenting density of the forest.

The Trammel's Trace crossing of the Sulphur was one of the most significant natural landmarks in the area. Anderson's Creek entered the Sulphur Fork a quarter mile above the crossing and deposited a shoal of silt and debris that created a convenient ford. Moscoso crossed here twice in 1542, and La Salle's surviving crew crossed in 1687.[33] In 1822, when immigrant traffic increased, there was no ferry to take travelers across the Sulphur. If they were lucky, the remains of some earlier crafted log rafts or pirogues were lashed to a tree on the bank of the river or pushed to higher ground by the last flood.

Trail rut leading to shoals across Sulphur River at site of Epperson's Ferry. Photo by author.

It was not until 1837 that Mark Epperson operated Epperson's Ferry at this crossing and it became a mail stop on one of the Texas Republic's early postal routes.[34]

The trail emerged on the other side of the Sulphur River bottom, on high ground between where Thomas Creek and Whatley Creek emptied into the river. The sharp rise of the trail to a bluff signaled an end to their present difficulties. Another long day of hard work through muddy ground and an arduous river crossing was complete. The next few days of travel beyond the Sulphur River offered a more relaxing journey with easier creek crossings. Thoughts of less demanding travels to come could now supersede the difficulties of the present journey. There was time for looking ahead but not for a good night's sleep. The insects did not allow for it.

Day 7: A Historic Fork in the Road

When rain clouds covered the sun in the early part of the day, weary travelers got a few more minutes of sleep before the sounds of thunder off to the northwest jolted them to attention. The scent of rain and stillness in the air alerted them to prepare for a day of wet travel. If rain continued until

Trammel's Trace ruts along Rusk/Panola county line filled by trees. Photo by author.

their next river crossing, they could only persevere or wait for the water to subside.

Anyone traveling on Highway 77 near Dalton Cemetery who knows where to look for the historical marker can still see ruts across what is now only a pasture.[35] A mile to the southwest, travelers came to a historic fork in Trammel's Trace. A trail from the north joined the main track from Fulton and continued southward.[36] This northern branch of Trammel's Trace connected the early Red River settlements of Jonesborough and Pecan Point to the main trail to Nacogdoches. The route between this fork and the Red River was later called the Spanish Trace.[37] Northwest of this fork, the trail to Pecan Point crossed the Sulphur River at what became known as Stephenson's Ferry, about one and a half miles upstream from where the Sulphur River crosses Highway 67 in the northwestern corner of what is now Cass County. The section of Trammel's Trace from Pecan Point and Jonesboro was the entryway for many immigrants from Tennessee and Kentucky, some of whom became part of Stephen F. Austin's "Old Three Hundred" original settlers.

Days 8–10: Turning South and East

Inevitably, the rains came. Not in torrents, but in a steady, light rainfall that sent the wagon underneath a canopy of trees until preparations to get back on the trail for a rain-soaked ride were completed. The next few days of travel were much simpler than the last. The creek crossings to come were more easily managed, and the sandy soil in the forests allowed the rain to soak into the ground beneath. Although the terrain was easier, the environs were not.

One hundred miles to the west of Trammel's Trace was a natural belt of trees and brush called the Cross Timbers, running north to south through Texas. Some believed this natural barrier was "fashioned by the Great Creator of all things to be a natural hedge between the woodlands of the settled portions of the United States from the open prairies which have ever been the home and hunting ground of the red man."[38] Running east of that brushy zone and curving to the southwest in the face of the tangled barrier, Trammel's Trace offered better cover, easier passage, and friendlier Indians. There was a chance of coming face to face with Indians who hunted and camped in the region. Reports of Indians varied so much that they were almost of no help at all. Some told exaggerated tales of bands of savage raiders stealing horses and guns and leaving their victims dead or defenseless in the wilds. Most travelers reported any encounters with Indians along the way as friendly. Caddo, Cherokee, Choctaw, and Delaware who hunted or camped in the area were counted among the most culturally advanced tribes. John Dunn Hunter, a notable Cherokee chief, would start out near Nacogdoches and hunt his way to the Red River, where he camped near Anglos who welcomed his presence.[39]

The Indians assessed the settlers' interests in remaining in their territory by trading with them. In spite of the general goodwill, immigrants new to the country could not help but wonder if the friendliness was merely a ruse before some horse thievery to be carried out during the night. The likelihood of a safe encounter and a little trade was not enough reassurance to dissuade the wary from paying closer attention to noises out of sight in these forests.

More certain than encounters with Indians, and perhaps more fearsome, was the frequency and predictably of stirring hornets and yellow jackets from their hiding places. The edges of creeks, the bottom sides of fallen logs, and tree stumps held swarms of the menacing insects. If a horse was stung, it would paw and kick the ground and then roll in the dirt to rid itself of the sting.[40]

From the fork in the road that was the Trammel's Trace branch to Pecan Point, the trail ran almost due south for about fifteen miles. There it passed near the site of an old Choctaw Village, just east of present Hughes Springs.[41] Near these settlements, a chalybeate spring filtered by iron ore reportedly had a healing effect on those who drank from the waters.

Leaving the old Choctaw Village, the path of Trammel's Trace turned southeasterly in a long sweeping curve, after which it followed a relatively straight course for about twenty-five miles to the crossing of Big Cypress Bayou, near present-day Jefferson. In one account of a trip back up Trammel's Trace from Nacogdoches to Pecan Point in 1821, the traveler said the only people he saw on the entire journey of almost 200 miles were one small group of Delaware Indians.[42] Caddo people built burial mounds in the area and would have left footpaths crossing the main trail. With signs of Indians about, even the sweet sound of a birdcall brought a certain uneasiness about whether it was really a bird or the signal of warriors waiting to attack.

A bluff at the edge of the Big Cypress bottomlands offered a vantage point high above the trail ahead. It was clear that there were few hills and little high ground for the next five miles. Halfway across the bottom, a single hill in the midst of the bottom was visible over the trees, another hundred feet higher than the bluff on which they stood. The trail headed in that general direction, and it was easy to see why. This change signaled travelers' entrance into an unhealthy country of river bottoms and thick hardwood forests different from anything they had encountered to this point. The woods were darker and more ominous, the high ground less certain, and the still air filled with more of the biting insects that tormented both people and horses. Mosquito bites left faces swollen, and buffalo gnats could kill a horse. The curdling cry of a panther, the howl of wolves in the distance, or the low moan of bears huddled in the canebrakes sent chills down the spine of the hardiest traveler. The prairies were behind them now, and terrain of a different sort lay ahead.

Days 11–13: Cypress Bayous

A lighter load made the muddy bayou crossing easier. Nothing was left on the wagon that was not needed. Food supplies dwindled after almost two weeks on the trail from Fulton. The need to hunt for food meant a gun was kept close by, as much for a sense of security as for hopes of dinner. There was still another week of travel before arriving in Nacogdoches.

The forests changed as the wagon headed down a slight incline into the expansive, wooded floodplain of the Big Cypress and Little Cypress Bayous.[43] The air was still, damp with the smell of rotting wood and wet earth. Magnificent cypress trees gave the forests a more ominous feel. The towering height of the old trees and the Spanish moss dangling from their branches created a haunting, uneasy atmosphere. The trail slogged across muddy sloughs from one dry stretch of ground to another for the next two miles. The horses made it through with equal amounts of prodding and reassurance despite their skittishness in the enclosed space. As range animals more comfortable in pasture, horses were easily spooked in the confines of the dense undergrowth, startled by noises in the shadows or simply by the tunnel of trees and brush. The wide base of each cypress tree was surrounded by dozens of woody roots called "knees" poking up through the black sloughs, hard protrusions that could injure horses and break down wagons. Fatigue wore down even the hardiest animals, and their wariness became more heightened as they grew tired. A horse laden with supplies was too wide where the trail was overgrown. Broken tree limbs and brush offered telltale signs of the prior passage of a wagon or a horse carrying pack bags. Travelers occasionally rediscovered the ruts of a trail used by others, but their advance for now was based less on markers of those gone before than on their sense of direction. They headed toward a solitary piece of high ground in the middle of the creek bottom and hoped for a short respite there. Even with all of the attendant obstacles, a crossing of the Big Cypress Bayou seemed routine at this point in the journey. Travelers learned to look for logjams backfilling a sandy island that bisected a faster flowing channel. They gathered up fallen logs and built upon the debris that was already there to construct an easier way to drive a wagon across a shallow ford.

The ground rose quickly out of the bottom not far beyond the Big Cypress Bayou. The change in elevation was so sudden that the trail curved around the rise to avoid the steepness of the grade. The top of the hill was almost two hundred feet above the swampy bayou behind them. From that vantage point, they could see their bottomland excursion had only reached its midpoint. The same type of terrain was ahead of them at the crossing of Little Cypress Bayou. The trail followed a slightly more elevated route through the bottom, and only the bayou crossing itself led to any difficulty. The Little Cypress was narrow but deeper than could be safely crossed in a loaded wagon. After another mile of travel, the trail finally emerged to higher ground out of the thick river bottom.

If providence and the trail builders were kind, a small clearing in the woods beyond the bottomland was available at about the same time weariness overcame tired bodies. Nightfall led to thoughts about how to respond if they spotted the last flames of a camp of Choctaw or Caddo sleeping near these same banks or suspicious characters taking cover in a canebrake. Enduring nightfall without the benefit of a fire for cooking or for light may have been advisable to avoid detection, but the risk of attack by humans was less than the sure assault of bugs. Biting insects were so numerous that the sound of their swarming drowned out everything but the shriek of a bobcat.[44] The smoke of a campfire stoked with sweet gum roots, called copal, was more valuable than any sense of security since it kept the bugs at bay.[45] The insects were far more likely to draw blood than panthers or Indians.

As the trail left the Little Cypress bottom, it wound up and down, from one small hill to the next, across narrow creeks for another full day of travel. The woods varied little, and the clearings were fewer. There were clear tracks around the largest trees and through the most open ground available. Two days past the mud and mosquitoes of the Cypress, the trail descended into another floodplain. Past Caddo burial mounds and across creeks and hills, the trail finally came to the edge of the last river crossing before completing the journey to Nacogdoches. The muddy, brown waters of the Sabine River lay ahead.

Days 14 & 15: Boundaries in the Piney Woods

The banks of any river became gathering places when flood waters prevented a crossing. Men of many nations camped together in a tenuous traveler's truce, their horses tethered nearby to feed and rest. When the water level allowed, the low water crossing of Trammel's Trace over the Sabine was atop an outcropping of dark brown lignite. The coal shelf was about forty feet wide and about three feet above the flow on its downstream side, creating a small waterfall. The shallow depth of the water and the hard surface made wagon crossings much easier any time the water level allowed.

The worst of the journey down Trammel's Trace was over. Unhitched from the wagon rig and tethered in a tall patch of cool grass on a moonlit night, the horses could eat until after everyone was asleep. Morning

Lignite outcropping at Trammel's Trace crossing of Sabine River near Ramsdale's Ferry. Photo by author.

brought preparations for the coming day with a vigor and optimism that had been subdued by the challenges of the trip thus far.

The path of Trammel's Trace angled steeply down the sandy bank of the Sabine River through the ruts left by previous crossings. The water there was slightly turbulent and formed a little waterfall over a shallow river at this spot. Black, flaky rock just beneath the surface, an outcropping of lignite coal, led straight across the Sabine to the other side. At low water, the crossing of the Sabine River at this rocky ford offered few challenges. The water was often only knee high at the edge of the coal shelf that formed the hard surface. The trail down to the river was steep and narrow, but once at the water's edge, the crossing proceeded uneventfully as long as the Sabine was kind.[46]

Upon emerging on the south side of the river, the traveler faced more of the predictable slog through bottomland filled with hardwood and brush. Although it was low and muddy, compared to the Cypress Bayou lowlands it was not nearly as difficult or as foreboding. Oxbow lakes, cut off from the former flow of the river, appeared on both sides of the trail. On the south side of the river, pines were more predominant and the character of the forests began to change. There was even more evidence of one of the demons of the Trace, poison ivy, in places where the canopy opened up to the sunlight. This poisonous plant grew widely as ground cover, plaguing travelers on foot who took to the trail in the wrong places. An itchy nuisance could become a serious infection without medicines or herbal remedies in harsh conditions unsuitable for cleanliness.

Keen-eyed travelers with more experience also noticed specimens of a tree that was one of the primary commodities of the Indians. Wood from the Osage orange, or Bois-d'arc tree, was easily polished, hard, and durable, making it a prime material for hunting bows.[47]

The trail south of the Sabine was wooded, but traveling was relatively easy—at least for the passengers. The horses pulled their load over and around hill after hill, weaving through the spaces between trees that were wide enough for the wagon. After crossing seven creeks and an overnight camp on a small summit, the next day travelers experienced more of the same kind of travel through the wooded canopy. Pines increased on top of the sandy hills; oak and walnut returned in the creek bottoms.

Just about where a second day past the Sabine crossing may have ended, a small, clear spring seeps out of a sandy bank on land presently owned by the author's family and bordering Trammel's Trace. The clarity of the water is enhanced by the pure, white sand that makes up the bottom of the shallow stream. Beneath the layers of topsoil, sand, and clay, less than a hundred feet below the surface, a layer of less permeable lignite coal slows the penetration of the ground water. As seepage hits the coal, it flows along the top of that black layer until it emerged along the edge of a sandy hill and begins to flow down the watershed and into the bayous and rivers. A similar exposure leads to a comparable flow in the opposite direction on the other side of the same hill. If one follows each of those insignificant trickles downstream, merging as they do with larger and stronger flows, one would find that the hill separates two watersheds. On the north side of the hill, waters from the spring flow to the Sabine. On the opposite side of the hill, the same waters flow south into the Attoyac and the Angelina. That uplift remains, and the old route of Trammel's Trace still passes.[48]

Days 16 & 17: One Ending Is the Next Beginning

The last week of travel down Trammel's Trace to Nacogdoches was through a virtually unbroken forest. Pine trees formed a dense crown overhead in the woods of eastern Texas. Short-leafed pine with a diameter of two to three feet dominated the landscape with a majestic evergreen shade, growing upwards of 130 feet. Loblolly pines were even larger, if less abundant. It would not have been unusual to see a loblolly with a diameter of four feet or more chest-high from the ground.[49] Trees like these were saplings over 200 years before Anglos used the trails that became Trammel's Trace. Trees like these had stories to tell.[50] Oaks like the chinquapin, red oak, white oak, and basket oak with acorn caps bigger than a musket ball were prevalent in the creek bottoms. Elm, black walnut, hickory, cypress, sycamore, dogwood, and birch provided a more varied forest in the lowlands. Other than tree limbs that naturally fell from the lowest parts of the pines, some sections of the forest floor were as clear as if cared for by unseen inhabitants. The only undergrowth was grass or lacy green ferns that thrived in the filtered sunlight. The size of the towering trees was a formidable deterrent for anyone who considered staking a land claim and building a cabin. Finding trees suitable for a cabin—small enough to handle with only an axe—was difficult in some stretches.

From a point just twenty-six miles north of Nacogdoches, the route of the old trail varied over time.[51] Williams Settlement was a well-populated community of over twenty families on high ground above the East Fork of the Angelina River, west of present-day Mt. Enterprise. Williams Settlement was established in the 1820s, before Texas was a republic, and was an active part

Through the Wilds 31

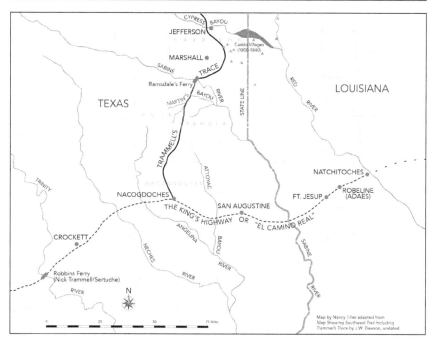

Southern half of Trammel's Trace. Map by Nancy Tiller, based on Dawson "Modern Map." Arkansas Historical Commission.

of the Mexican district Nacogdoches. A road from there to Nacogdoches would have been established in the early days of the settlement for trade and business activity. Most likely, the road through Williams Settlement was Trammel's Trace. Farther south and west on Dill Creek, just north of present-day Cushing, Texas, the site of the Spanish Mission San José de los Nazonis was served by trails to nearby Indian settlements. The mission was established in 1716, one hundred years before the beginnings of Trammel's Trace. Spanish and French trade goods recovered during archeological studies of the mission site attest to early activity predating the use of Trammel's Trace.[52] Any remaining segments of pathways from the mission to Nacogdoches may have provided a part of the trail that became Trammel's Trace.

Indications of the likely route of Trammel's Trace between Williams Settlement and Nacogdoches also appear on a map for a land survey just south of the current boundary of Rusk and Nacogdoches counties. In the survey for Mariano Sanchez, a road from Pecan Point to Nacogdoches—Trammel's Trace—tracked diagonally across the survey to the south-southeast, toward Nacogdoches from the direction of Williams Settlement. The plat map for the Luis Sanchez survey, angling off to the west of Trammel's Trace, names a "road from the Saline to Pecan Point."⁵³ The "saline" was a salt deposit northwest of Nacogdoches that was a critical resource for the region. That road connected to the El Camino Real near the Trinity River, a key crossing.

The last stretch of Trammel's Trace into Nacogdoches followed the eastern edge of La Banita Creek—a sandy trail along a beautiful creek. Signs of an approaching arrival in the only settlement in this part of Texas could be recognized in a slight widening of the road or by the presence of small cabins. Finally, Trammel's Trace became El Calle del Norte (North Street) in Nacogdoches and intersected the El Camino Real near the old stone building at the plaza. Nacogdoches was virtually empty after 1812, its inhabitants driven off by conflicts with the Spaniards. After 1820, some of the early residents returned to their homes, bringing the population of the settlement back to over one hundred inhabitants. Nacogdoches and its trading partner Natchitoches, to the east in the United States, were both important to the history and commerce of the borderlands. Natchitoches was a river port and the outer reaches of US influence. Nacogdoches was a way station on the El Camino Real, which led westward to the new colonies in Texas.

The rigors of travel and the strain of uncertainty never diminished the dreams that kept immigrants moving forward into Texas, following promises of land, riches, and opportunity. Small reminders of those dreams were found along the way south. A surprise bounty of peaches and grapes planted long ago by the early inhabitants offered an unexpected trailside reward. Stands of old-growth forests with oak and pine hundreds of years old and a circumference that three grown men could not reach around provided a tangible reminder of the sense of infinite possibility they sought. The trip from north to south on Trammel's Trace, from Fulton, Arkansas, all the way to Nacogdoches, Texas, took up almost three weeks of travel over 180 miles of raw, wagon-busting trail. Those who managed to keep their supplies and belongings dry past dozens of stream crossings and three rivers, and who avoided mortal injury, disease, and attack, at last reached their next waypoint in Nacogdoches.

As they emerged from the forest at the place where Trammel's Trace widened into something closer to a road, the need to be on guard did not disappear; it simply changed to an awareness of new dangers. Any traveler coming down the way of the smugglers might be suspected to be another one of the bad men the trail tended to bring south. Seeing someone emerge from the road's shadows on the way south gave the citizens of Nacogdoches cause to wonder what act of illegality was underway. That assessment was probably correct in the earliest days of the trail's use, before the wave of colonization and migration that started so fervently after Mexico took possession of the territory in 1821.

There were two jumping off points on the northern ends of Trammel's Trace heading to the new country to the south—Fulton and the route from Jonesborough and Pecan Point. On the southern end was Nacogdoches, the town that signaled it was time to turn west to a better life toward colonization in central Texas. In between were smugglers, thieves, Indians, and the men and women brave enough to endure those risks in order to start a new life in Texas. Those inhospitable wilds and unsettling times were the province of Trammel's Trace.

3

The Trammells of Kentucky and Tennessee

At length my father gave me a young race-horse, which well-nigh proved my everlasting ruin.

—PETER CARTWRIGHT, PREACHER
IN LOGAN COUNTY, KY[1]

Old roads and trails were commonly named for where they led. The California Trail, Santa Fe Trail, and Oregon Trail were early roads that led to the destinations named. Of the thirty old trails designated as part of the National and Scenic and Historic Trails system, only three are named for people. The explorers Lewis and Clark, Capt. John Smith of early Plymouth, and a Spanish commander, Juan Bautista de Anza, have trails designated by their names.[2] Even the Wilderness Road across Cumberland Gap, scouted by the widely known Daniel Boone, does not bear his famous name.

Within only a few years after traders and immigrants began using it frequently, Trammel's Trace was named for Nicholas Trammell.[3] Nick Trammell was a farmer and trader from the Kentucky-Tennessee border who moved to the Missouri Territory in 1807. Like many others, he migrated south to Missouri and Arkansas, then on to early settlements on the Red River along what is now the Texas-Oklahoma border, and then farther south into Texas. At some point between 1813 and 1821, Trammell had cleared enough of the existing Indian trails or begun using the old trails for trade and horse smuggling so much that his name identified the route. For his contemporaries to name the trail Trammel's Trace meant Nicholas Trammell and his path were inseparable in the public consciousness. It meant no other name made sense. It was Trammel's Trace in the possessive. It was Trammell's by virtue of his regular use of it for his smuggling enterprise and because he opened it up to others who used it either for their own illicit intent or for immigration to Texas.

Moving west along early roads into emerging territories was in the Trammell family genes: from England to Stafford County, Virginia, from Virginia to Tennessee and Kentucky, and from there to the Missouri and Arkansas Territories. Members of Trammell's family served as scouts during the Revolutionary War and had seen the beauty and bounty of land west of the Appalachian Mountains. The Trammell's movement toward places where governments were not yet in place was a consistent aspect of the family history. Their record of early involvement and key roles in organizing frontier settlements put them at the head of the line when migration created new opportunities for the family businesses of salt making, gambling, and trade.

Tennessee and Kentucky

Nicholas Trammell, the only son of Nicholas Trammell Sr. and Frances (Fanny) Maulding, was born in 1780 near present-day Nashville, Tennessee. In the same year, his mother's family, the Mauldings, settled just north of what became the boundary line between Tennessee and Kentucky.[4] The Trammell and Maulding families were both from Virginia and likely met there before their migration west. Travel from Virginia to the Tennessee-Kentucky border was by boat along the Holston River or by way of riverside trails into east Tennessee. Across the Cumberland Gap on the Wilderness Road, early settlers moved south and west to the Kentucky Road.[5]

Long hunters were adventuresome men who traveled from North Carolina to western Tennessee in search of game. They visited the area twenty years before Nicholas Trammell was born and were likely the first Anglos to set foot on the land where Nashville now stands.[6] These hunting parties made their way to nearby creeks, where they built "station camps" to serve as a base for hunting expeditions. They were there to hunt the buffalo and deer attracted to countless salt licks in the region. Even flocks of parakeets, now extinct, migrated there for the minerals found in the saline-laced earth and springs.[8]

Station camps were rough structures covered with bark or hides used as base camps. They could be as small as eight by ten feet. One side was open

"The Station Camp-Dogs and Deerskins", by David Wright. Courtesy David Wright.[7]

to the fire for warmth, and the other sides were enclosed or backed up to a rock or large tree. Eighteen miles from a salt lick along the north fork of the Red River (of Tennessee), the Mauldings of Virginia built a small collection of similar structures that became known as Maulding's Station.[9] It was along the road to Danville, eighteen miles from Kaspar Mansker's station camp, one of only three in the region by 1783.[10] Even before the station camps were established, Morton Maulding, one of Nicholas Trammell's uncles, hunted the area around the salt licks: "I think in May [1780], I went hunting from Manscoes [Mansker's] Lick, and fell into a buffalo trace at the head of Muddy River at the Big Spring and pursued it to the middle Lick at which place I cut the first letter of my name and encamped all night. Upon my return I gave information of the Lick I had found and no person known of them, I name one Moates Lick by which name it has been called ever since."[11]

The Trammells and Mauldings were not only among the first to arrive in these Tennessee settlements, they were also actively engaged in the formation of the first local governments. In May 1780, Nicholas Trammell Sr. and two of Fanny Maulding's brothers were signatories to the Cumberland

Compact. The Compact codified articles of self-governance and land use. The 250 men who signed the compact created an orderly life for themselves and in the process founded the city of Nashborough.[12] This early settlement, once known as French Lick and now the city of Nashville, was surrounded by Indians unhappy about the intrusion on their lands. That unrest and the senior Trammell's tenacious nature resulted in tragedy for his family.

Father Killed

Cherokee Indians killed Nicholas Trammell's twenty-six-year-old father late in 1783 or early in 1784.[13] Trammell Sr. and Philip Mason were hunting along the headwaters of White's Creek, a few miles north of Nashborough. A group of Indians opened fire on them while the pair was skinning a deer. The Indians wounded Mason, stole the deer, and left the area. Trammell responded to the theft of the deer carcass by going to get reinforcements from nearby Eaton's Station. Trammell, Mason, and four others found the trail of the venison thieves and followed them. What the pursuers failed to notice was that one by one the Indians left the main trail to hide in the woods. When the members of Trammell's search party believed they could ambush a couple of the remaining attackers, Indians hidden to their rear came upon them from behind. Mason was mortally wounded, and the other settlers ran away quickly. Passionate in his desire for revenge, Trammell rebuked the other members of his party for their cowardly retreat. Since they were outnumbered, the settlers left and rode back toward the station camp.

Trammell's anger was still simmering when he came upon a man named Josiah Hoskins. His contemporaries described Hoskins as a "soldier braver than Julius Caesar, and an excellent rifleman."[14] Trammell told Hoskins about the events and made his case for another attack, this time with Hoskins's gun rekindling their bravery. After tracking down the Indians, Trammell and Hoskins restarted the fight and quickly killed three of the Indians. When Trammell and Hoskins came out of hiding and stood together to fire another round, each was shot and died instantly. The rest of the Anglo settlers held their ground and kept up the battle, but both parties soon became exhausted and uninterested in the fight. They ended the shooting by common consent. Like boxers fighting to a draw, both sides were bloodied but not beaten.[15] An early account of these events led one author to compose a poem.

> They had killed and were skinning a deer;
> They thought not that Indians and danger were near;
> For they were both brave men, who never knew fear;
> And laughing and talking, kept skinning the deer.
> The Indians, they heard them, and slyly crept near,
> And fired down upon them, not front, but in rear,
> And wounded Phil. Mason, not badly, 'tis clear,
> For he fled, with Nic. Trammel, half-way to the station,
> While Nic. hastened on to 'rouse the whole nation.
> Phil. tied up his wound, looked sharp for the foe,
> And for this friend Trammel and others, who'd go,
> And fight to the utmost for the fat doe.[16]

After Nicholas Trammell Sr.'s death, the court in Davidson County, Tennessee, recognized his wife, Frances, as administrator of the estate on January 7, 1784. Her brother, Ambrose Maulding, also appeared in court to support her.[17] The next day, the court appointed Capt. James McFadden and Solomon White to appraise the estate and return an inventory to the court at its next quarterly session.[18] When Trammell's estate was finally assessed six months later, in June 1784, the inventory suggested his young family lived on bare essentials. His estate consisted of two cows with calves, two small steers, a featherbed, two pewter plates, one iron pot, and two steel traps.[19] In addition to these basics of life, the estate included three horses. A bay mare and colt were for working and riding, but an Eagle mare was the type of horse that could race for a purse—an early indication of the source of young Nicholas's future interests.

The sparseness of the Trammells' belongings is difficult to understand, even for a frontier family. The parents of Nicholas Trammell Sr., Phillip Trammell and Jemima Grimes Trammell, were landholders with hundreds of acres in Logan County, Kentucky. Their land and that of their remaining children was part of what was known as the "Georgia Settlement" spanning the border of Kentucky and Tennessee.[20] Fanny Trammell's Maulding relatives held early positions of some prominence and also owned personal property, slaves, and land.[21] The young Trammell family is interesting in its contrast. That it took six months and three court terms to inventory and appraise so short a list may simply be a consequence of the pace of frontier justice. The reason for the delay is documented in the court record. At the April 1784 session of the court, months after Trammell's death, Ambrose Maulding testified that James McFadden was prevented from returning an estate inventory due to his sickness. The court allowed him to report at the next session in June 1784, and the estate was finally fully transferred to Fanny.[22]

1792—Changing Boundaries and Families

Little is known about Fanny Trammell and her young son, Nicholas, in the years immediately following her husband's death in 1783. With her family's roots north of Nashborough near Maulding's Station, it could be expected that Fanny Trammell relocated there after her husband's death, nearer the family support she needed. Nine years later, in 1792, when she and her son reentered the historical record, the Maulding and Trammell families were impacted by significant changes.

The summer of 1792 was a busy one for the creation of governments and jurisdictional boundaries in the region. Kentucky separated from Virginia and became a state on June 1, 1792. Logan County, Kentucky, home base for Fanny Trammell and her Maulding kin, became the thirteenth county formed in the new state by September of the same year. The first Logan County court met in the home of Richard Maulding, Fanny Trammell's brother. It was there that her brother Ambrose was elected justice of the peace and that another brother, Wesley, was appointed sheriff. Sheriff Wesley Maulding did not take long to get in full swing with his job. By 1793 he complained to the county court that the jail was not sufficiently large to hold all of the citizens needing incarceration.[23] In the same year, the new governor appointed a third brother, Morton Maulding, to be a major in the militia. The prominence of the Mauldings also led to acquaintances with those who would soon have influence beyond the region. The Trammells and Mauldings most certainly knew the young lawyer Andrew Jackson, a future president of the United States, who practiced law from the Logan County courthouse as early as 1794.

The first road officially marked and opened in the newly formed Logan County crossed Ambrose Maulding's creek and headed toward Nashborough, running between the Logan County log courthouse and the line between Kentucky and Tennessee.[24] This road was likely the same one used by the Trammells and Mauldings many times to travel between their family's settlements.

There were other changes on the horizon for the Trammell family beside the events going on in the territory around them. Sometime before 1792, Nicholas Trammell's widowed mother married Zachariah Askey from Logan County, Kentucky.[25] Although the date of their marriage is uncertain, there are records of another significant change in the family that same year. In 1792, Frances Maulding Trammell Askey transferred guardianship of her son, twelve-year-old Nicholas, to the Trammell side of his family.[26] His father's brother, Phillip Trammell, would raise young Nicholas Trammell.[27]

In today's social climate, we might assume that young Nick was unhappy with his new stepfather or that an adolescent rebellion led to conflict in the family. Given the nature of the Trammells' and Askeys' continued migration west together, a more likely possibility is that the Trammell kin simply convinced Nick's mother to allow them to raise him in the way of his father's family—trading, salt making, and leading the way into new territory. Since both the Askeys and Trammells later migrated to the Missouri Territory and Nick's new half brother, Morton (Mote) Askey, was allied with Nick for life, it is unlikely that there was any serious family dispute.

Religion in "Rogues Harbor"

Despite the prevalence of the Mauldings in local government, or perhaps as a result of it, Logan County, Kentucky, was known for its lawlessness and gambling. Peter Cartwright, a notable American Methodist Episcopal preacher, lived in Logan County from 1793 until 1802 and was a contemporary of young Nicholas Trammell. In his autobiography, Cartwright said that as a young man in Logan County he was lost to the sins of gambling at cards and horse racing.[28] "At length my father gave me a young racehorse, which well-nigh proved my everlasting ruin; and he bought me a pack of cards, and I was a very successful young gambler; and though I was not initiated into the tricks of regular gamblers, yet I was very successful in winning money. This practice was very fascinating, and became a special besetting sin to me; so that, for a boy, I was very much captivated by it."[29] Although his father provided him with the instruments of his "special besetting sin," his mother's prayers apparently held sway. At age sixteen, Cartwright received religion and developed a different perspective on his upbringing in Logan County: "Logan County, when my father moved into it (1793), was called 'Rogue's Harbor.' Here many refugees from all parts of the Union fled to escape punishment or justice; for although there was law, yet it could not be executed, and it was a desperate state of society. Murderers, horse-thieves, highway robbers, and counterfeiters fled there, until they combined and actually formed a majority. The honest and civil part of the citizens would prosecute these wretched banditti, but they would swear each other clear; and they really put all law at defiance. . . . This was a very desperate state of things."[30]

This view of the arrangements between those who would commit criminal acts and those who would overlook them could be taken as a direct indictment of Trammell's kin, the Mauldings. Since one of Sheriff Wesley Maulding's first official statements was to ask for more jail space, it appears that the Mauldings may have opposed the legal discrepancies described by Cartwright. Of course, if the sheriff and the justice of the peace were brothers, legal matters could certainly be more easily decided—or ignored.

Although Trammell had a family history of gambling and racing horses that was similar to Peter Cartwright's, there is no evidence that Nicholas Trammell's encounters with the courts in Kentucky involved murder, robbery, or other violent acts. Trammell was five years Cartwright's senior, however, and certainly could have been qualified to initiate Cartwright into, as he called them, the "tricks of regular gamblers."

In a place named Rogue's Harbor, the public definition of illegality must be kept in perspective. Another account from Cartwright does that. Cartwright was sent to boarding school at the home of Dr. Beverly Allen, a Logan County physician who was once a traveling preacher in Georgia for the Methodist Episcopal Church. While in Georgia, Allen reportedly "fell into sin, violated the laws of the country, and a writ was issued for his apprehension."[31] When his apprehension approached, the good Reverend-Doctor Allen apparently did not wish to be apprehended and warned the sheriff not to enter his room or he would be shot. When the sheriff persisted and attempted to take Allen into custody, Allen shot and killed him. After Reverend Allen committed this crime in Georgia, he fled to Rogue's Harbor in Logan County, Kentucky, where the Trammells and Mauldings made their home. That Cartwright's parents felt he might find redemption from his gambling at the home of a murderer who escaped justice from another state is indicative of the general view of illegalities and jurisdictional authority at the time. By comparison to his earlier troubles, Allen's home was apparently considered a safe place for a young man's education by Logan County standards.

Other men of religion attested to the deplorable moral condition of Kentucky and Tennessee at the time. A brief description of the character of the people summed it up succinctly: "Politically they were violent and dogmatic; morally they were corrupting; and, in respect of religion, they were utterly infidel."[32]

Into that environment came James McGready, a Presbyterian minister. McGready was instrumental in lighting the fires of a Great Revival of 1800 that began near where the Trammells, Mauldings, and Askeys made their home. In July 1800, the first "camp meeting" in Logan County set souls afire across the entire region. The Great Revival of 1800 began in Virginia but spread instantly across the country. McGready advertised the meeting broadly and invited residents from over a hundred miles away to come to the edge of a prairie for four days of preaching and salvation. The camp was a hollow square with a covered stand in the center and wagons or makeshift shelters lining the boundaries. Barton W. Stone, a Presbyterian minister, was present at the meeting and described the fervor: "The scene was new to me and passing strange. It baffled description. Many, very many, fell down as men slain in battle, and continued for hours together in an apparently breathless and motionless state, sometimes for a few moments reviving and exhibiting symptoms of life by a deep groan or piercing shriek, or by a prayer for mercy fervently uttered. After lying there for hours they obtained deliverance."[33]

So where was the twenty-year-old Nicholas Trammell while this eventful massing of the newly faithful was taking place? What was his reaction to the waves of religious change sweeping the territory? If the answer to those questions was available, it might do much to explain the enigma of Trammell's personality, behavior, and motivations. What influence the revival had on him is uncertain, but the actions that the Trammells and Askeys took in the years following are clear—they soon left Kentucky behind for opportunities farther west.

Troubles in Kentucky

As the Mauldings became entrenched in Logan County politics, Nicholas Trammell prepared to leave the region and move west with his Trammell and Askey kin into an almost unimaginable expansion of US territory. On December 20, 1803, the United States doubled in size. President Thomas Jefferson hoped to convince the French to cede New Orleans to protect the mouth of the Mississippi River and defend the United States. In the process, Jefferson got the entire Louisiana Territory—800,000 square miles of land. With one stroke of the pen, the United States acquired an enormous tract of land for future expansion and settlement and assured the free navigation of the Mississippi from the interior down to New Orleans. When

Napoleon sold Louisiana to the United States in 1803, there was no agreement in effect that clearly defined the boundary between Spanish Texas and French Louisiana. Language in the agreement specified Louisiana was transferred "with the same extent that it now has with the hands of Spain, and that it had when France possessed it."[34] That lack of clarity on the boundary contributed to years of disagreement and allowed a borderland culture of smuggling, theft, and lawlessness.

The Trammell and Askey families began considering a move into the newly acquired territory not long after the Louisiana Purchase in 1803. In the years between 1807 and 1810, Nicholas Trammell and his extended family made their way between Livingston County, Kentucky (formed from part of Logan County) and the Missouri Territory west of the Mississippi River to what is now northeastern Arkansas.[35] Although their movements seemed deliberate and planned, legal and financial troubles slowed their pace.

1803—Lawsuit in Kentucky for Slander

Nicholas Trammell's history of court appearances across his entire adult life will demonstrate that he took matters personally when he believed he had been cheated or lost an advantage. Rarely did he accept the judgment of a single court unless it was in his favor. Only weeks before the grand acquisition of territory in the Louisiana Purchase, Nicholas Trammell was focused on defending his name. On September 8, 1803, Trammell accused a man named Peter Fletcher and his wife of slander.[36] Although Trammell was the plaintiff in the suit, his stepfather, Zachariah Askey, appeared in Trammell's place in the matter, acting as his "next friend" to the court. Trammell was not in Livingston County at the time of the hearing—he was perhaps already looking for new opportunities in the Missouri Territory.

Zachariah Askey had his own legal dispute with Fletcher two years earlier, in March 1801, over an unspecified matter. Askey lost the original case. The court dismissed Askey's appeal and forced him to pay court costs.[37] Askey may have been eager to support Trammell in this matter against Fletcher since the outcome of his own suit left unsettled issues with the same opponent.

The parties to Trammell's lawsuit were finally in court on September 8, 1803. Fletcher pled not guilty and the case was continued to another date. Robert Montgomery was summoned to testify on Trammell's behalf and was likely related to a man named James Montgomery, who had also sued Fletcher for slander at the same time as Trammell. The circuit court jury finally heard the case on March 7, 1804, but no judgment was rendered. As often happened with the passage of time, the parties got over their initial disagreements. On September 4, 1804, Nicholas Trammell agreed to dismiss the suit. Based on Trammell's character, he probably agreed only after Fletcher and his wife apologized or paid in some way or after Trammell at least felt vindicated. Nicholas Trammell was not one to be slighted or treated unfairly.

1803, October—Failure to Pay Taxes

Like many others who headed west to avoid debts or criminal charges, Trammell was feeling pressure from the legal system in Kentucky in circumstances that may have encouraged his relocation to the Missouri Territory. Nicholas Trammell and his stepfather, Zachariah Askey, were both called to task by the Livingston County, Kentucky, commissioner of tax in October 1803.[38] Trammell had not provided his list of taxable property to the county, even though he was previously cited by the court for that failure. Zachariah Askey went one step further and openly stated his refusal to provide a list of his property. As a result of their actions, the court summoned Trammell and Askey to appear and explain why they should not be triple taxed as a penalty for their noncompliance. Tax records that immediately follow this action show each of them with only small tracts of land and a few horses. Since they owed the county money, we can only wonder if the court confiscated the eight shillings, about $1.25, Trammell earned in November 1803 as a bounty for a wolf hide.[39]

1804—Claims for Land in Kentucky

Unclaimed land could still be found in Kentucky, but more often than not ownership of the land was difficult to determine rather than unassigned. Court records are filled with listings of hopeful land claims. In 1804, Nicholas Trammell was trying to secure land in Livingston County. He entered a claim for fifty acres on May 8, 1804 and was apparently successful.[40] The 1805 tax list shows him with fifty acres and four horses on the Ohio River.[41]

Nicholas Trammell's half brother, Mote Askey, recorded claim #1171 for four hundred acres in October 1805 in a similar fashion. Askey withdrew the claim in November when he discovered that the land was owned by someone else.[42] Zachariah Askey had five hundred acres on the Ohio River surveyed and patented in 1805, but the 1808 tax list showed him as owner of only fifty acres.[43] By 1807, tax records indicated that Nicholas Trammell and Mote Askey no longer owned land in Livingston County, Kentucky, likely sold in preparation for a move to the Missouri Territory. Only Zachariah Askey still had land on the tax rolls in Kentucky. None of the three appeared to have been slaveholders while they were in Kentucky. Zachariah's two horses, Mote's two horses, and Trammell's six horses were their only taxable property.[44]

1806, September—Mote Accused of Felony in Livingston County

Nicholas Trammell and Zachariah Askey were not the only ones with legal problems as a result of tax evasion. In 1806, Mote Askey was listed on the Livingston County delinquent tax list as "gone to Tennessee."[45] He may have in fact gone to the Missouri Territory to examine the unsettled lands around the White River. His disappearance was not permanent, however.

In June 1806, the Livingston County Circuit Court returned a felony indictment against Mote Askey for stealing a heifer "with force and arms" from Israel Stilley (Stilly) on May 20, 1806. Mote was forced to appear in Livingston County Circuit Court to answer to a charge of felony theft.[46] On July 15, the sheriff was instructed to arrest Mote and bring him before the court on its first day of session in September 1806. By late August, seven witnesses were called to appear and describe the larceny. Mote appeared in court with his father, Zachariah, on September 2, 1806, to post bond of $250. Mote agreed not to leave the jurisdiction of the court, and the case was continued until March 1807.

In spite of the bond, Mote did not appear in court in March, which was a serious mistake. He was rescued from immediate apprehension when James Trimble, John Young, and Samuel Bruton offered a bond for his appearance at the May 1807 term of the court. Jurors were called for Askey's trial on May 21, 1807, and Wilson Montgomery was called to testify. Before the facts of the case could be presented, the judge ruled *nolle prosequi*, meaning that the prosecutor chose not to continue the case against Mote Askey.[47] With seemingly simple cases taking months and sometimes years to resolve, this type of result was common. Mote Askey escaped prosecution.

Looking West

The Trammells and Askeys had had enough of Kentucky. Whether drawn by opportunity or driven out by too much attention from the courts, a pivotal year for Nicholas Trammell and his Askey kin followed Mote's troubles in court. By 1807, Nicholas Trammell began establishing a settler's claim to a square mile of public land along the White River near present-day Batesville, Arkansas, in what was then the Missouri Territory.[48] But before leaving Kentucky for good, Trammell had land business that required years of court battles to secure. Trammell's tenacious nature in legal matters would be evident once again.

1807, August—Land Warrant #33

Nicholas Trammell inherited rights to claim two 640-acre grants of land made to his father as a result of his military and militia service. It appears that Trammell did not learn of these claims until he was an adult. After he discovered the land claims, Trammell began a history of disputes surrounding them that connected him to the most notable attorneys in two states and demonstrated his persistence in pursuing legal remedies. From his new home in the Missouri Territory, Trammell made several trips back to Tennessee and Kentucky in attempt to secure his claims, a process that took almost twenty years.

One of these grants came as a result of his father's service as a soldier in Captain Benjamin Logan's Kentucky Company during the Revolutionary War.[49] Nick's mother, likely with the help of her Maulding brothers, filed preemption certificate #33 under this grant for land near Nashborough in 1785, little more than a year following her first husband's death.[50] After his mother died, Nicholas Trammell inherited the two land claims.[51] Although his mother laid a claim to the land near Nashborough, it had never been patented and secured. Consequently, in August 1807 Nicholas entered warrant #33 on his own, but not for the land his mother claimed along Station Camp Creek near her home. Trammell's new claim was for land along the Elk River in south central Tennessee.[52] The land was surveyed on October 16, 1807, and in June 1808, the warrant was granted. Nicholas Trammell became the owner of 640 acres in middle Tennessee at the same time he was relocating to the Missouri Territory.

1807—Certificate #816

Additionally, as a reward for Nicholas Trammell Sr.'s "heroic defense of Davidson County" in his fatal battle with the Cherokee, Fanny Trammell received a second grant for 640 acres of land in 1787. She filed preemption certificate #816 five years later in August 1792 for land more specifically identified as "along upper fork of Station Camp creek about three miles from a small Sulphur Lick under the ridge running about three hundred yards from a small spring marked A.M. running down the branch including the Spring."[53]

Trammell's uncle, Ambrose Maulding, marked this spring with his initials, as his brother, Morton Maulding, had done earlier with the salt lick he discovered. The ridge is a notable geologic feature that runs through Sumner County, Tennessee, from northeast to southwest. The hollows between ridges were the source of numerous springs feeding Station Camp Creek.[54]

The timing of the second preemption claim #816 suggests some questions about whether it was made with young Trammell's best interests in mind. Nicholas Trammell's mother transferred guardianship for him one month prior to the date her family members staked a land claim using his father's second certificate. Although the claim was marked, it was never settled or patented. Zachariah Askey and Richard Maulding, Trammell's stepfather and uncle, employed a prominent attorney and land speculator, John Overton, to secure this preemption claim in September 1800, probably just after Fanny's death. Askey and Maulding made an agreement with Overton to give him half of the 640 acres in return for his efforts to secure the dormant grant.[55] At the time of the claim in 1792, Nicholas Trammell was only twelve years old—not yet at the age of majority. It was not until 1807, the same year Trammell staked his claim on land in the Missouri Territory, that claim #816 surfaced again.

As Trammell was cutting his ties to Kentucky, he learned that this second 640-acre claim was his as heir to his father, but it was not secured. Trammell sued Overton in 1807 to break the agreement previously made by Zachariah Askey and Richard Maulding. Trammell was already in the Missouri Territory, so he appointed another relative, Ragland Langston, to secure the warrant on his behalf. Since Langston was part of Trammell's extended family, he contracted to get the work done on more favorable terms than giving up the 320 acres Overton was to receive.[56] Trammell's suit alleged Overton got wind of Trammell's initiative through his network of land dealing and then beat Langston to the land office and removed the certificate, preventing Langston from completing the assignment. In court, Trammell agreed that Overton located and surveyed this second claim on the Elk River, in White County (now Franklin County, Tennessee), but that he did so only when he learned Trammell hired Langston to do the work Overton had neglected. The basis of Trammell's lawsuit was that the agreement with Overton was made with Askey and Maulding, not with him, and it was made while he was a minor. Trammell felt he should not be bound to that agreement as an adult and sued to avoid being forced to hand over 320 acres to the land-rich Overton. This case would not be settled quickly.

1807—Improvements on Land in Missouri

Trammell's urgency in settling his land matters may have been spurred by his relocation from Kentucky to the White River in the Missouri Territory. When Trammell discovered the two claims in Tennessee, he realized that they were an opportunity to make a sale and fund his removal to Missouri. Trading along the White River was beginning to show promise due to the influx of Anglo inhabitants and a confluence of available means of transportation. The White River area was already inhabited by Indians adapted to the presence of Anglos, mostly Cherokees, who traded horses and pelts for cooking utensils and cloth. One of Nicholas Trammell's uncles built his trading post among the friendly Cherokee in 1807. John Lafferty was also a trader there and a contemporary of Nick Trammell's father from the Tennessee side of the Kentucky-Tennessee border region.[57]

The Trammells and Laffertys, on new land in the Missouri Territory, were in a prime location to carry on frontier trade. Their posts were at the intersection of a river navigable for hundreds of miles and the only road from the territories in the east to the emerging settlements farther south and west. This strategic location allowed access to two means of moving trade goods across the region. The White River was navigable even during low water. It was bordered by a prairie 27 miles wide and 150 miles long, running from its headwaters all the way to the Arkansas River.[58] The south side of the White River bordered Oil Trough Bottom, a fertile and productive forest about fifteen miles long and three miles wide filled with huge oaks and canebrakes. Bears hibernated in the thick brush during the winter, their stomachs full of muscadines and paw paws. The bears attracted hunters, both Anglo and Indian, who boiled the meat for oil and used the skins for trade. Trappers and hunters collected and preserved great quantities of beaver, otter, deer, raccoon, and bear skins during the summer and fall. They then took their pelts down the White River to the Arkansas Post in canoes, where they met merchants coming upriver with salt, iron pots, axes, blankets, knives, and rifles for trade. Arkansas Post was a government-sanctioned trading post that opened around 1805.

1809, June 28—Trammell in Jail

Although he was settling in the Missouri Territory, Trammell still traveled between there and Kentucky. One of his trips back to Kentucky did not end favorably. On June 28, 1809, Nicholas Trammell, described in the court records as a "yeoman" farmer, found himself behind bars in Livingston County, Kentucky—jailed for twelve days for an unspecified felony.[59] "Nicholas Trammell who stands Indicted by the grand Jury for felony was brought into court and on motion of the Defendant by his attorney its ordered that this prosecution be continued until next and the said prisoner is remanded to Jail."[60]

Trammell never ignored an opportunity to respond when he believed that he had been cheated or belittled. He had a history of using the courts not only to secure property or debt but to make a point. After sitting in jail for twelve days, Trammell was released on a recognizance bond. The prosecutor secured his witnesses against Trammell with bonds requiring their appearance for the September term of the court. Edward Lacey Sr., William C. Rodgers, Edward Head, Hugh Carothers, and George V. Lusk provided bonds of $200 each, and Elijah Battes (Bates, Battas) provided a bond of $1,000.

As soon as Trammell had an opportunity to speak to the court, he turned the tables on the legal system. Trammell filed a complaint with the circuit court against the jailer, David Kline. Kline was known as much for his transgressions and lawsuits against him for debts as for his possession of the jailhouse keys. Trammell claimed that while he was jailed, Kline sought a $1.12 payment in return for the "favour and ease" Kline showed him while he was a prisoner. Before the court even made a decision on Trammell's guilt of the unspecified crime for which he was jailed, Kline was indicted for his abuse of a poor prisoner. Based on the court's findings in the matter, it appears Trammell asked Kline to make a purchase for him of 75 cents. Then Kline demanded a bribe of 50 percent, an additional 37½ cents beyond the purchase price of the unnamed goods he agreed to deliver to Trammell while he was behind bars. Trammell was in no position to argue and paid the extra money, but Kline wound up with an indictment for accepting money beyond that which he was due. On August 22, 1809, the circuit court directed Kline to appear and answer to the charges, and he was found guilty a month later. Kline was fined 37½ cents by the court and ordered to repay Trammell the full $1.12 he received from him. Although Kline was chastised for accepting what was not legally his, shortly after this judgment the court paid Kline the money he was legitimately owed for keeping Trammell as a prisoner.

> For arresting the body of Nicholas Trammell charged with felony $2.50
> Per account filed ordered that David Kline for committing and releasing Nicholas Trammell on a charge of felony 3 times $1.25
> For victualing [feeding] the same 12 days $3.37[61]

Trammell may have avenged his mistreatment by Kline but was ultimately called before the court to answer to his felony indictment on Wednesday, September 27, 1809.[62] The case was surprisingly continued, perhaps for lack of witnesses, and bond was increased to a total of $3,000. The size of the bond was significant and probably based more on the expectation that Trammell might leave the state and return to the Missouri Territory than on the nature of the unknown crime. The return court date was set for March 1810, but Trammel already had one foot out of the state. His return to appear in court seemed unlikely.

1809, November—Appoints Maulding and Pitts to Represent

Nicholas Trammell was ready to be more than just another farmer battling over land that was becoming increasingly settled. In the Missouri Territory he could be a trader who took risks for commercial gain. After his jailing in June and court appearance in late September 1809, Trammell was ready to move on by November of that year. He appointed Morton Maulding and Bartley Pitts as his agents to transact business for him regarding the disputed land claims with Overton. This document was also signed by his wife, Sarah, the first indication of Nicholas Trammell's marriage.[63] As Trammell left Kentucky for the Missouri Territory, he left both criminal and civil matters unresolved. Either his patience for those troubles wore thin or the opportunities in Missouri were more certain.

Trammell's departure may have had to do with a different kind of opportunity as well. A man with the last name of Trammell found gold north of Little Rock in 1809. The newspaper account provided only a last name, so there is no way to know if this was Nicholas, a member of his family, or an unrelated man with the same last name. This Trammell found gold while hunting near the foot of Crystal Hill, a high bluff on the north bank of the Arkansas River about fifteen miles above Little Rock. The unnamed Trammell showed the ore to Frederic Notrebe, the key trader at Arkansas Post, and learned that it was indeed gold worth over one hundred dollars. The news spread quickly and caused "great excitement among the bold and adventurous." An expedition was soon organized at New Orleans to go in search of gold.[64]

1810—Failure to Appear in Court

Trammell had had enough of Kentucky. Though he had been moving between the Missouri Territory and Livingston County for about two years, in 1809 he finally decided to cut his ties and move on. The 1809 tax lists for Livingston County, Kentucky, show no record for Nicholas Trammell. He was gone.

Trammell was still under indictment for a unnamed felony, and a $3,000 recognizance bond stating that he would "not depart without leave of the court" was still in effect. On March 26, 1810, Nicholas Trammell's name was "solemnly called" in the Livingston County Circuit Court, but he did not appear to answer to the charges.[65] The court forfeited his bond and set his case for the June term of the court. His case file does not reflect his appearance before the court until September 27, 1810, when he was "led to the bar in custody of the sheriff."[66] Trammell pled not guilty and a trial was held the same day. The jury agreed with the plea and Trammell won yet another protracted victory, outlasting the legal process.

The next day, the prosecuting attorney for the Commonwealth of Kentucky dismissed the case. Trammell's criminal obligations in Kentucky ended, but his civil disputes there continued with the land he accused Overton of taking. While Nicholas Trammell was refocusing his life in the Missouri Territory, events farther south in Spanish Texas unfolded on a separate path, one with which he would soon connect.

4

1800–1812
Boundaries under Pressure

> . . . *notwithstanding my repeated orders, some extractions of horses are still made to Louisiana, it may indeed be by secret roads.*
>
> —GOV. J. B. ELGUEZABAL[1]

Smuggling Horses and Contraband

Trade was the center of economic activity. Native people engaged in long-distance trade with partners hundreds of miles away long before the Spanish came to the New World. Later, when Spanish and French traders sought them out, Indians offered pottery, *bois d'arc* bows, buffalo hides, salt, and corn in trade for hard goods and arms. The Indians wanted not only trading partners but to establish a kinship through the exchange of gifts. That kinship was a key element in their ventures and broadened their culture of trade and commerce.[2]

The Spanish approach to developing relationships was to establish contact through religious missions rather than trade. In general, Indians did not respond favorably to the Spanish, giving the commercially minded French a significant advantage. The French understood the importance of kinship, but the Spanish did not. The Spanish commandant at Nacogdoches in 1780, Antonio Gil Y'Barbo, was forced by his superiors to withdraw his promises of trade and gifts to the Indians as a result of French dominance of trade in the region. The otherwise friendly tribes viewed this act as a dismissal, leaving Y'Barbo and the Spaniards "entre la espada y la pared"—between the sword and the wall.³ The wall was the official barrier to conducting trade across the Sabine, hopelessly imposed by Spanish authorities, and the sword was the military authority of the Spanish crown. The commandant of Nacogdoches noted in 1806 that friendly Indians were trading more with the French and Americans in Natchitoches inside the expanded US territory than with the Spanish in Nacogdoches. He recognized that the Indians lacked the necessary commitments to the Spaniards and that trade in Natchitoches was more advantageous for the Indians because they received more gifts from the French.⁴

The ability to establish trading relationships was not the only French advantage. A French trader also developed many of the early trading trails across East Texas. Louis Juchereau de St. Denis was responsible for the transformation of trails through woodlands and prairies into well-worn trading paths. St. Denis was credited with that effort by his Indian trading partners. In 1783, a Kichai Indian chief said it was St. Denis who had "first opened the trails in all their nations and who had made the peace."⁵ In fact, St. Denis's relationship with the Indians across northwestern Louisiana and northeastern Texas may have been largely responsible for French trading relationships with the Indians. The strength of his presence, through his descendants, led to strong trading partnerships for three generations.⁶

Events in Spanish Texas during the early 1800s led directly to Trammell's future opportunities for trade and smuggling in Texas and to the emergence of Trammel's Trace. Historic trading relationships and the existence of trails for the purpose of trade created conditions that were irresistible for those bold enough to exploit them. As the nineteenth century began, uncertainty over the boundary between Spain and the United States triggered two decades of illegal trade and smuggling in one of Spanish Texas' most sought-after resources—wild mustangs that roamed the prairies along the Red River. Capturing wild mustangs in large numbers on the open range was a difficult and dangerous operation. Anglo traders generally left that part of the enterprise to others. They preferred to trade goods from the United States for horses captured or stolen by someone else, most often by Indians. Some mustangers learned the Spanish method for capturing horses. Although that method was violent and often brutal, the final outcome was that thousands of wild stallions, mares, and colts were captured and sold in markets in the United States across the Sabine River in Natchitoches, Natchez, and New Orleans.

The Spanish means of capture was to construct out of brush and natural cover a large pen with a funneled entrance. They chased huge herds of wild horses toward the entrance and corralled them in the hidden pen. When the horses discovered they were trapped, two to three hundred at a time, the wildness of the captured herd often resulted in horses being trampled and killed in the chaos. After the best horses were selected and culled, the rest were set free and the dead left for the vultures. The captors deprived the stock of food, prevented them from taking any rest, and continually kept them in motion, actions that finally broke the wild horses to submit to the saddle and bridle.[7]

Anglos sometimes used a method that resulted in as many deaths as captures. Creasing a horse to bring it down relied on the improbability of an exceptionally accurate shot at the base of a horse's neck that nicked the spine, temporarily paralyzed the horse, and allowed it to be subdued. Shooting a fleeing mustang at a spot just behind the ears or in front of the hips requires expert marksmanship, as well as luck, even at close proximity. Some horses were lucky. In 1807, Indians found a small bay horse among mustangs on the Sabine River with a hole the size of a rifle ball clean through its neck near the windpipe.[8] The likelihood of missing the mark was very high, and a missed shot often killed the horse. When mustangs were abundant or inexpensive due to a glut in the markets, the mustangers could not have cared less.

Creasing horses courtesy of The Portal to Texas History. "'Creasing' Mustangs in Texas." Star of the Republic Museum, Washington, Tex. http://texashistory.unt.edu/ark:/67531/metapth31782/

1801—Philip Nolan

Spanish authorities had bigger political concerns than the minor leakage of trade goods or horses across the border. They were up in arms about the smuggling carried on for years by Philip Nolan. Nolan, a mustanger and filibuster from Ireland by way of Kentucky, made the first of many trips into Spanish Texas in 1791. His trips across the border to capture and return horses by way of the El Camino Real through Natchitoches to markets in the United States resulted in a brisk traffic. Nolan carried out his mustanging ventures in spite of the fact that such trade was illegal in the eyes of the Spanish. Even if the authorities granted traders passports to legally enter Spanish territory, their nonpayment of royal duties made their efforts an illegal repudiation of the Crown's distant authority.

On Nolan's first incursion into Spanish Texas in 1791 his trade goods were confiscated. He remained in Texas to live with Indians for the next two years. During that time he learned even more about the herds of horses running the prairies and the Spanish and Indian ways to capture and break them. After Nolan returned to New Orleans, he made a second expedition across the border into Texas at the end of 1795. He returned to Natchez, Mississippi, with 250 head of horses. Mustangs regularly sold for $50 a head, with the best going for up to $150.[9]

It was not just Nolan's horses and the money to be made that created intrigue. Nolan arrived in Natchez accompanied by Andrew Ellicott, boundary commissioner for the United States. The prospect of having an official representative of the United States across the border in Spanish Texas placed suspicion on Nolan's motives and intent. Nolan's trips also caught the attention of one man uniquely focused on the expansion and development of the United States. On June 24, 1798, Thomas Jefferson wrote a letter to Nolan and requested that he provide information on the mustangs that ran wild across the territory. Jefferson's letter described scientific and historic interests, but many were convinced that Nolan's trips to Texas were part of Jefferson's plan to gain the territory from Spain. Nolan appeared to admit his purposes in a letter he wrote in 1800: "Everyone thinks that I go to catch wild horses, but you know that I have long since been tired of wild horses."[10] Nolan prepared to meet with Thomas Jefferson to discuss the wild horses, and probably wild lands, but no record of a meeting exists. Instead, Nolan undertook his most successful expedition back in Texas. He returned to Natchez at the end of 1799 with more than one thousand two hundred horses, undeterred by Spanish authority. One can only imagine the difficulties of that journey, moving over a thousand animals cross-country.

Natchez was the southern end of the Natchez Trace, which originated near Nashville. News may have traveled slowly, but it certainly would have been news in Tennessee and Kentucky if that many horses arrived in Natchez from Spanish Texas. The trading Trammells would have no doubt heard about these cross-border opportunities for trade and smuggling and considered how they might be a part of such a grand venture.

Philip Nolan left Natchez on his final journey into Texas at the head of a body of well-armed men in October 1800. Carrying a double-barreled shotgun, a pair of pistols, and a carbine, Nolan made his way to the area north of Nacogdoches, where he encountered the Indian trail that became Trammel's Trace.[11] Communication among Spanish officials the following month indicated that Nolan's activities were taken quite seriously by the authorities. Directions for his arrest and fears of his resistance were strongly communicated to the commandant in Nacogdoches. Spanish troops sent from there intercepted Nolan and his party while he was corralling mustangs in northeastern Texas. After a brief skirmish, Nolan was killed on March 21, 1801. A surviving member of his party, Peter Ellis Bean, crossed paths with Nicholas Trammell almost twenty-five years later.

Eliminating Nolan did little to deter trade across the border between Spain and the United States. Stories of the money he made from trading expeditions into Texas led others to look for opportunities beyond the legal boundaries of the United States. In May 1801, Spanish officers in Nacogdoches communicated their concerns about "the introduction into that Province of prohibited goods, and on those of the removal of horses toward the territories of Louisiana."[12] Clandestine extractions of droves of horses continued to be carried out over the road from La Bahia (Goliad) to the border, the lower road of the El Camino Real. The governor acknowledged what he and others knew but could do little about: "notwithstanding my repeated orders, some extractions of horses are still made to Louisiana, it may indeed be by secret roads, or by virtue of contracts which the interested parties conclude with the citizens of that post [Nacogdoches] in their dwellings."[13] Smuggling contraband and horses was an ever-present and pervasive reality in the borderlands between Spanish Texas and the United States. If Spain was truly concerned with stopping smuggling, it sorely underestimated the number of Spanish subjects interested in illegally trading goods for horses.

1803—Smuggling at Nacogdoches

The official position of the Spanish government on smuggling in early Texas was that it was an illegal activity that soldiers must prevent. Military patrols, correspondence, and pronouncements all supported the stance that citizens of the Texas province must direct their commerce westward to San Antonio and stop trade across the shorter distance to Louisiana in the United States. The practical reality was that smuggling across the Sabine River to the east better met the economic needs of the Spanish citizens in the Nacogdoches District. When faced with the option of purchasing readily accessible goods in Louisiana or hundreds of miles away in San Antonio de Bexar, as they were expected to do by Spanish authorities, residents made the convenient choice to trade to the east despite the illegality. When the residents of Nacogdoches complained to Nemesio Salcedo, the commandant general of the Interior Provinces, about not being allowed to acquire goods from across the river in the United States, Salcedo recommended harsh and decisive action on the part of the authorities in Nacogdoches: "fall back on the punishment or confiscation of any goods, fruits or manufactures, that will be introduced into Nacogdoches proceeding from Louisiana, since, notwithstanding the greatest distances and highest prices, the inhabitants of it have to provide themselves from the other villages of that Province, by not being permitted to do so from a foreign colony."[14]

From time to time, Spanish soldiers at Nacogdoches actually accomplished their mission of subverting illegal smuggling and capturing the perpetrators. Forty-eight horses and mules were seized in March 1803, the result of a smuggling attempt by a mulatto named Dennis.[15] These minor victories, although celebrated to the glory of Spain, did little to stop the waves of smuggling and trade across the border. Governing from a distance, the Spanish authorities could make all the pronouncements and proclamations they wanted to try to restrict illegal trade. They could send soldiers, change commandants, or guard roads and crossings, but the inevitability of trade across eastern Texas would not be slowed.

Smuggling was not a solitary venture. Anyone who could herd horses, travel cross-country, ease river crossings, quietly move trade goods, or be on the lookout for Spanish soldiers scouting the area was a mutually beneficial co-conspirator. At the key crossing of the El Camino Real at the Trinity River, ninety miles west of Nacogdoches, the Spanish government discovered its edict against smuggling had little impact. An Anglo named Charles

Boyles was permitted to operate a ferry on the Trinity for the benefit of those traveling between Nacogdoches and San Antonio de Bexar. The commandant of Nacogdoches in 1800, Jose Miguel del Moral, alerted the governor that Boyles was allowing "the secret introduction of prohibited goods by river, from New Orleans or Natchez."[16] Even after Boyles was removed from his post, the Spanish were losing their battle against smuggling. The governor was perhaps unaware his own officials were corrupt in governance and adept in aiding the smugglers. Market forces were much more powerful than the reach of the Spanish Crown.

1803—Fortifying Nacogdoches

After the Louisiana Purchase became part of the United States in 1803, protection of the new borderland between Spanish Texas and the United States became an area of concern for both countries. The Spanish military sought to increase its influence in Texas in an attempt to regain lost authority. They posted regular patrols on the road from San Antonio to Nacogdoches to discourage entry into Texas by traders and smugglers from the east. Spain made some improvements to the El Camino Real but not for the betterment of settlers and criminals. The existing roads were prepared in case of war with the United States and to provide better access to the region in the event that troops needed to move quickly. Surprisingly, it is unlikely any of these improvements were made on the path to the north that became Trammel's Trace. Spanish concerns were primarily focused on the United States to the east and San Antonio to the west. Land to the north of Nacogdoches was inhabited by Indians, and Spain mistakenly viewed them as a buffer to any migration or trade from that direction.

Migration pressures from the north existed beyond the piney woods and Indian villages of eastern Texas and increased as a result of actions taken by the US government. In 1803, the Congress of the United States allocated money to improve the Southwest Trail across the Missouri Territory from St. Louis to what is now Fulton, Arkansas. The use of the military to hack underbrush meant travel heading south would be simpler. With the stated intent of the improvement to support military purposes, those efforts led to a more common reference to the trail as the Congress Road, National Road, or Military Road. With freshly cut trails making travel south less difficult, the number of Americans settled at the edges of the young nation increased. There they could wait and watch for even more opportunity across the uncertain border toward Spanish Texas.

Map of southwest trail across Arkansas. Map by Nancy Tiller, based on Dawson map. Arkansas Historical Commission.

1804—Solutions and Supporters

The death of Phillip Nolan may have been a short-term deterrent for smuggling, but the deterrence did not last. By May 1804, another party of US traders left Natchitoches on horseback headed toward the upper Red River. José Ignacio Y'Barbo led a military contingent from Nacogdoches with orders to stop them, but the trading party could not be found.[17] Spanish authorities in the Nacogdoches settlement continued to fight illegal trade, with little success. Virtually every resident between the Trinity and Sabine Rivers was suspect in this commerce of necessity and opportunity. The Spanish were forced to think of new ways to reassert their control and remove the smugglers from the borderlands. "The greater part of the citizens of this place are scattered on ranches which extend almost to the Sabine River contiguous to the settlers of Bayou Pierre, who are a little further on. Most of them are concealers of contraband goods. It might not be a bad idea to remove them from this said section so that, in this way, the crimes as well as the thefts which are constantly committed may be stopped."[18]

The military commandant of Nacogdoches, José Joaquín Ugarte, suggested his forces move settlers from the east of Nacogdoches to the west side, away from the unguarded proximity of trade routes to the United

States. There they could be more easily monitored and controlled. Ugarte hoped he could placate the settlers by compensating them with legally assigned land. Perhaps he even convinced himself for a short time that this approach could work. Given the accessibility of secret roads leading from the Trinity River around the post of Nacogdoches, this change was unlikely to help even if the settlers agreed to move. Ugarte's only other idea was to issue a proclamation to make smugglers afraid of being caught and jailed, a plan with little chance for success. These were not people inclined to be fearful of authority, and his plan was never carried out.

One of the options Spain exercised to try to control the illicit trade was to license only certain parties to deal in imported goods with the Indians. William Barr was one of those purveyors. Spanish agreements with Barr allowed him a sanctioned trade monopoly with Indian tribes to the north of Nacogdoches, the region later crossed by Trammel's Trace, and to the west toward the Trinity River. Barr had already demonstrated his drive for control of trade and the high price he would exact to maintain it. When Phillip Nolan's removal of mustangs threatened Barr's trading territory, Barr joined Spanish soldiers in killing Nolan and his party. Spanish authorities interpreted this as Barr's loyalty to the crown, but it was more likely Barr's commitment to his own interests. In 1804, Barr requested that the Spanish authorities prohibit anyone other than his traders and their servants from introducing goods into the region. Barr also asked that the unauthorized sale of brandy to the Indians be halted. Barr demanded that he be able to continue to acquire goods from New Orleans, in complete opposition to the Spanish edict for the citizens of the Nacogdoches District. Barr's request to bypass San Antonio and turn to the United States acknowledged the same practical, economic demand for goods that the authorities were trying to subvert.

The most blatant of Barr's requests was that he be allowed to continue to export horses he acquired through trade with the Indians, horses the Indians had either captured or stolen from other tribes. His request added the caveat that he would sell them only in the Spanish colony of Florida rather than in the United States, bypassing the nearer market in New Orleans.[19] Although his offer may have been beyond any credible belief, Spanish authorities acquiesced to every request Barr made. Their condition was that any goods coming from New Orleans be used solely for trade with the Indians and not for trade with Anglos or other settlers. Barr's commitment was an unlikely one: Spanish authorities did not have the manpower to monitor Barr's compliance. In addition, there were substantial numbers of unlicensed actors carrying out the very same trade without Spanish sanction, a fact that was part of Barr's complaint. As a result of the agreement, Barr was able to accomplish legally what Spain deemed illegal for others. It can be safely assumed that in the name of commercial gain, Barr exceeded any limits Spain sought to place on his business dealings and no doubt traded with anyone who could pay the price.

1806—Trinidad de Salcedo

Smugglers always seemed to find a way around, but that did not prevent the military from going to great lengths to prevent illegal trade. In an attempt to outflank the smugglers, Spanish authorities founded the military post of Santisima Trinidad de Salcedo in January 1806 on the east side of the Trinity River at the El Camino Real.[20] This key location was an important crossing where goods brought upriver could be transported along one of several clandestine routes that bypassed Nacogdoches into the United States. One of those paths went near an important salt works and continued northward on to Trammel's Trace.

Poor road conditions were a limitation on the military's ability to patrol the region. Governor Saucedo wrote the commandant of Nacogdoches in January 1807 about problems with the conditions of the roads and the difficulty of river crossings between Trinidad and Nacogdoches. Saucedo directed that soldiers be kept not only at the Trinity crossing but also at the crossings of the Neches and Angelina Rivers along the El Camino Real toward Nacogdoches. Troops were to guard the canoes and ferryboats and prevent them from being used by anyone involved in suspected commerce. In addition to Spanish protection for safe crossing of these three rivers, Saucedo instructed the commandant to make certain that anyone bearing dispatches or mail along the El Camino Real be a good swimmer.[21]

Through various creeks and miry places, across the Mustang Prairie east of Trinidad, amidst cane reeds and swamps, the soldiers patrolled in search of smugglers by day and slept little by night. River crossings took three to four hours, a frequently documented testimony to the difficulty of the routes. Horses were exhausted from slogging through muddy paths. Conditions like these were less tolerable to those trying to enforce the ban on contraband than to those willing to endure them for personal gain.

The United States across the River
1806—Freeman-Custis Expedition

In the summer of 1806, Spanish troops at Trinidad, Nacogdoches, and all across Texas received a call to arms for something more troubling than the movement of trade goods. This time it was not smugglers or thieves but an official expedition of the United States testing Spain's ability to control its northern border along the Red River.

As a result of the amount of activity on the roads between Nacogdoches and Natchitoches, Spanish attention was heavily focused on the Sabine River border with the United States to the east. Smuggling and contraband trade were pervasive and required constant attention by the authorities. Since the Red River was a great distance to the north and there were no settlements between it and Nacogdoches, Spain devoted much less attention to the ill-defined reaches of their northern border. That negligence changed suddenly in April 1806.

Only three years after dispatching Lewis and Clark to explore the west, Thomas Jefferson sent Thomas Freeman and Peter Custis on an expedition up the Red River from Natchitoches. Freeman was a civil engineer and surveyor, and Peter Custis was the first academically trained naturalist to conduct a major scientific probe into the American West. Jefferson had three goals in sending the Freeman-Custis exploratory force up the Red River: to test reports that the Red River might provide a commercially viable watercourse to Santa Fe; to entice the region's Indians to align with the Americans; and to probe the Louisiana Purchase's disputed border with New Spain. Although it was billed as a scientific survey, the Freeman-Custis expedition had strong commercial and political overtones. By July 1806, the expedition drew the attention of Spain and "threw their whole country into commotion."[22]

Spain mustered a large percentage of its force in Texas in response. At the end of 1805, there were only 141 Spanish troops in Nacogdoches and 40 in Trinidad. Only seven hundred Spanish troops were stationed across all of Texas, a testament to the underestimation of the force required to maintain control of the province.[23] When the Spanish military heard about the expedition, troops were dispatched from other parts of Texas to Nacogdoches to protect the border, and 250 soldiers quickly moved up the trails of East Texas toward the Red River. A patrol that size needed a road to make its way north, and if there was not a road, they made one. Sections of the trail they blazed later became a part of Trammel's Trace (Spanish Trace) from a point near present-day Naples toward the Red River.[24]

As the American expedition rounded the Great Bend of the Red River, Spanish forces moved to a point where they could intercept the Americans. At the base of a hundred-foot-high bluff on the Red River, Spanish troops made their force visible to the approaching Americans.[25] Journals from the expedition noted 150 Spanish horses on the beach in the initial confrontation. The Spanish captain in charge of the force, Francisco Viana, explained his orders to Freeman. He was there to prevent any US force from venturing upriver before the boundaries were defined. Freeman asserted that his purpose was to simply explore the river, but Viana was undeterred. President Jefferson's instructions were to withdraw if challenged, so the US expedition of the Red River ended on July 30, 1806, turned back by Spanish troops. Although this challenge at the northern border of Spanish Texas was averted, only three months after Freeman and Custis were turned back, another military confrontation seemed imminent on the Sabine River east of Nacogdoches.

1806—Neutral Ground Agreement

With the boundary of the Louisiana Purchase unsettled, posturing and negotiations between the United States and Spain continued for years after the huge acquisition. President Thomas Jefferson asserted that all of Texas was part of the Louisiana Purchase, but Spain resisted that assessment. Spain defined the boundary at the Arroyo Hondo, a rather small stream between the Sabine River and Natchitoches. To declare their claim and protect their border, in 1806 Spain posted additional troops at the Sabine and in Nacogdoches. US soldiers responded in similar fashion along the Arroyo Hondo, creating the potential for armed conflict. Gen. James Wilkinson, commander of the US armies in the area, proposed a demilitarized zone where neither country's troops would patrol from the west bank of the Sabine River to Arroyo Hondo on the east and the area stretching undefined from there to the north.[26] Wilkinson's proposal was not a treaty or even a diplomatic resolution of governments; it was simply an accord between military leaders until a more permanent boundary agreement could be defined. In Wilkinson's letter to his Spanish counterpart, he stated he was "willing to risque the approbations of my Government to perpetuate the tranquility of the inhospitable wilds."[27] Later evidence revealed that Wilkinson was in fact an agent of Spain and was risking more than mere retribution.

The Neutral Ground agreement consisted entirely of correspondence between Wilkinson and Gen. Simón de Herrera of Spain, exchanged on October 29 and November 4, 1806. With Herrera in Nacogdoches and Wilkinson in Natchitoches, their representatives met somewhere in between and exchanged the proposals. Even though the officers found a way to avoid an escalating conflict, a result of the agreement was a significant strip of land on the edge of each country's frontier completely devoid of military influence or control. The Neutral Ground became a free trade zone for lawlessness for years after the agreement. Criminals, squatters, exiles, traders, and a host of undesirable others were drawn to this area where they could operate unfettered by authority. Spanish accounts by visitors to Nacogdoches and the Neutral Ground describe a territory filled with contraband trade on both sides of the boundary. Only a few years after much of the borderland was deemed a no man's land, Trammel's Trace crossed right through the fringes of that ungoverned land.

The Spanish Response
1808/1810–Smugglers Captured near Trinidad

Market forces, roundabout roads, and complicity on the part of Spanish officials appointed to prevent smuggling were major barriers to the Spanish goal of eliminating clandestine trade. Even Philip Nolan's killing did not deter smugglers with similar goals. Anthony Glass, another Red River mustanger, followed the market for inexpensive Texas mustangs Nolan had exploited. Glass and ten others successfully conducted trade and captured horses during a ten-month trip to North and Central Texas in July 1808.

Spanish troops not only patrolled the borderlands and river crossings but also attempted to impede the smugglers in more passive ways. The military constructed log barricades on the banks of the Trinity River at a well-used crossing known as the "second falls," a natural crossing downstream from Trinidad de Salcedo.[28] The barricades must have been a laughable impediment to resourceful smugglers unfazed by such a limited blockade. Soldiers did occasionally win a minor victory, intercepting smugglers in the act. Soldiers patrolling from Trinidad in 1808 caught Henry Quirk, Joseph McGee, and Joseph Brenton in possession of 159 horses and mules and three jackasses, a significant amount of four-legged commerce to move across the back roads.[29] When the captured smugglers were thoroughly questioned, they implicated the commandant of Trinidad, Don Pedro Lopez Prieto, as consenting to or tolerating the exportation of horses through that post. As was often the case, the men in charge of preventing smuggling were involved in its success.

Besides the human obstacles to stopping illegal trade, there were natural limitations on the effective presence of Spanish soldiers. If the military established stationary posts, smugglers simply found another river crossing or blazed another trail to avoid detection. Spanish authorities were acutely aware of contraband trade bypassing Trinidad on alternate routes toward and around Nacogdoches. A network of roads south of Nacogdoches that came to be called *El Camino del Caballo*—the Road of the Horse—was already well used for smuggling goods.[30] In January 1810, the commandant's log recorded the assignment of a patrol in search of secret roads: "one sergeant, one corporal and nine soldiers patrolled the roads, paths, and side-roads by which contrabands of goods or herds of horses may be able to be carried toward foreign dominions; and in case of meeting up with any, to arrest them."[31]

In this marketplace of noncompliance, Indians were partners as well as customers, informants for the Spanish as often as they were guides for the Anglos. In May 1810, sixteen Americans with merchandise to sell asked Tahuayases (Taovaya) Indians to steal horses for them in trade for goods from the United States. In the same prairies where Philip Nolan was killed, Indian allies of Spain reported eight Americans and two Spaniards had chased mustangs without any interruption. Although the Indians seemed willing to inform the Spanish soldiers to gain their good graces, they were unwilling to join forces with them to eliminate the Anglo smugglers. A Spanish corporal's letter to his captain noted the Indians' reluctance to engage the Americans crossing the border illegally: "these [Indians] were not going with ours [soldiers] to arrest them, on account of being deprived of ammunition and the Americans are well-armed, and of a disposition to die rather than surrender."[32] The Indians knew the Anglos were more committed to continue the illegal trade than the Spanish soldiers were to prevent it.

The post at Trinidad continued to be the focal point for Spanish military efforts to stop the movement of stolen horses to the United States and the entry of unauthorized traders from the east. The Spanish wanted to believe an additional post upriver from Trinidad would help in spite of the evidence to the contrary: "In order to avoid any people may cross by the crossings above of this river, who may come from Bexar toward Nacogdoches, or from Nacogdoches toward Bexar, without presenting themselves in this town [Trinidad], with what they may bring, and cut off their passports, I have determined, for the present, and until consulting with your Lordship, to keep there three detached men, in order that they may scout daily, and detain any that may attempt to make passage by those crossings. . . ."[33] Posting three reluctant solders at a strategic crossing was a paltry response. If the Spanish believed the armed smugglers would fight to the death, then these three soldiers far from home in a strange wilderness must not have slept well at night. Spain's response to its own rhetoric frequently came up short.

The acquiescence of those in charge of preventing smuggling was also an ongoing concern for the Spanish. Bishop Marin de Porras visited the mission of Nacogdoches in 1805 and found that Governor Herrera not only had been complicit in allowing unsanctioned trade but in fact benefited from it personally. In an 1809 letter, the bishop recalled the ongoing corruption as a warning for present officials, saying that Herrera "dealt equally in another very lucrative contraband trade in mules and horses, with which the Americans have enriched themselves."[34] The military commandant at Nacogdoches also displayed either a lack of information or a willingness to deceive. José María Guadiana informed Governor Salcedo in 1810 that the citizens of Nacogdoches did not trade with anyone and in fact lacked all initiative for commerce.[35] Rather than acknowledge the citizenry's complete abandonment of Spanish law, he recommended pity on the poor villagers as a more appropriate response. Guadiana already knew that Spanish efforts to patrol, barricade crossings, and post soldiers along the main roads were an exercise in futility. His purported view was that the involvement of those in his jurisdiction in smuggling contraband and other vices was a moral issue, the result of idleness born of poverty. In 1810, Governor Salcedo visited Trinidad and Nacogdoches hoping to see the problems himself. He traveled for three weeks down El Camino Real in a coach drawn by twenty-four mules.[36] Journal accounts of his three-month stay focused more on the Indians who came to pay homage to him than on any high-level attention committed to reduce the scope of the problem.[37] Guadiana, as commandant, and Salcedo, as governor, either could not or would not take adequate measures to stop the smuggling and trade.

1810—Crow's Ferry on the Sabine

The ferry crossing at Trinidad was a strategic location west of Nacogdoches at the edge of Mustang Prairie near the Trinity River. Not only did the main routes of the El Camino Real converge there, but smugglers' roads around Nacogdoches also had to emerge from secrecy and cross the river at one of the shoals north or south of the legitimate ferry crossing. Trinidad was the first line of interior defense against the exportation of horses coming from the interior, but in order to prevent incoming contraband trade, Spain pushed its presence as far east as the Neutral Ground agreement would allow—the Sabine River.

Spain still controlled the Sabine River crossings of the El Camino Real on the west side of the river after the Neutral Ground agreement. There were multiple crossings of the El Camino Real at the Sabine River, each used at different times depending on conditions. Jose Miguel Crow, an English farmer and an early settler in Nacogdoches, was licensed by the authorities to operate a ferry at the lower or "Royal Crossing," where Spanish troops maintained a post.[38] Crow moved his ferry upriver, out of the view of Spanish troops, after they caused him to lose two of his flatboats. Crow wrote the governor a letter in 1810 filled with complaints, resentful of the agreement that abandoned him on the edge of an ungoverned territory. He told the governor that he could not control either the smuggling or the fact that people from the United States were settling in the borderlands: "Being out of all Jurisdiction of any Laws as a place of aselem as City of Refuge and I want to know what I am to do in this case as you have put the Land out of your Power and threw me a way also where I can neither have Law nor justis from your Excellency."[39]

Crow went on to explain that perhaps the same informants who had shown ill will toward him to the governor were involved in theft and contraband trade and should be suspects themselves. Crow believed his worst enemies were in Nacogdoches and was concerned enough about his letter falling into the wrong hands that he asked the governor to burn it. That the letter survived is probably an indication that there was much in Crow himself that the governor did not honor.

1810—Factories and Indian Trade

In addition to the contraband that flowed into Texas from the commercial centers of Natchitoches, Natchez, and New Orleans, trade goods also moved upriver from those settlements toward traders and immigrants gathering in the Missouri and Arkansas Territories. The US government also sought to sanction and control trade but with a more resourceful and realistic approach than Spain. The government opened a sanctioned trading post, called a factory, at Arkansas Post up the Ouachita River in 1805 in an attempt to control the growth of unlicensed traders and to build relationships with Indians in the region. Factories were government-controlled trading posts intended to monitor and control trade with the Indians and to force unlicensed traders to move farther into the frontier. Rather than simply pulling in trade from Indians already in the Missouri Territory, the increased trade at Arkansas Post attracted tribes east of the Mississippi River who were pressured by the United States to move farther west.

Pressures to move west resulted in a confluence of cultures and a competition for land. A mixed-blood chief named Connetoo, also known as John Hill, formed a village in what is now northeastern Arkansas as early as 1796 and expected six hundred Cherokee to join him.[40] Cherokee Chief Sanlowee (Tsulawi or Fox) and Chief Duwali (the Bowl) began their migration to the Missouri Territory down the Tennessee River in twelve canoes and a flat-bottomed boat in January 1810. Both Connetoo and Bowl lived in the same part of the Missouri Territory as Nicholas Trammell. Bowl and his tribe later migrated to Texas during the same time frame and within the same geographic region as Trammell, making it likely they had a trading relationship. Chief Connetoo and Nicholas Trammell soon crossed paths in a trading dispute that began Trammell's reputation for lawlessness.

Even with the backing of the US government, Arkansas Post could not survive competition with unlicensed traders. The factory there closed in 1810, leaving unsanctioned traders like Nicholas Trammell to garner the business in furs, horses, and hard goods. Trammell took trade matters into his own hands. On November 10, 1811, he executed a note to John Young for the purchase of twenty pounds of beaver fur. The following June 5, Young's home near Arkansas Post was broken into and the note was stolen, making it impossible for Young to collect payment. Nicholas Trammell was not charged with the crime but was certainly believed to be responsible. It would not be the last time his name was associated with shady business dealings.

1811—Sibley's Report

Pressures from every boundary with the United States continued to impinge on Spanish sovereignty and expose their inability to control the borders, but it was not only pressures from outside Texas that weakened their authority. The citizens of Spanish Nacogdoches were already calling themselves Americans by early 1811 and openly advocating commerce across the river. In a letter on December 31, 1811, John Sibley, the US Indian agent stationed in Natchitoches, noted that trade interest from within Spanish Texas added to the tension: "There is Considerable Trade Carried on from the Spanish Country (Texas) to This Town (Natchitoches), dollars are brought in Packed on Mules and the Same mules are Packed out with Merchandize, Waggons, & Carts are likewise used."[41]

The Neutral Ground created a haven for the worst kinds of people—people with more than a passing interest in taking advantage of those who traveled the El Camino Real. Sibley asked for US military protection for the citizens in the area not from the Spanish but from the Anglo outlaws who were committing various crimes in the region. Congress responded to these conditions and proposed a joint patrol of the Neutral Ground by Spain and the United States to eradicate the outlaws from the borderlands. Spain refused to accept any changes to the original Neutral Ground agreement, so the smuggling and contraband trade continued unabated.

The combination of isolation and lack of authority led to many opportunities for criminal activity. Sibley reported the robbery of Spanish gentlemen who came to Natchitoches from San Antonio with $20,000 in cash to conduct their trade. On the way back, they were waylaid and robbed by outlaws from the United States who lurked in the Neutral Ground. The plundering criminals were later killed and the merchandise retaken.[42] Businessmen in Natchitoches in 1812 complained to the United States of the robberies.

> The commission of robberies on the Territory West of the Arroyo Hondo, and East of the Sabine River, and as your memorialists believe within the limits of your Excellency's Government, has become so frequent, that it is no longer safe to travel on the highways & roads through which the commerce of this Parish with the adjoining Mexican [Spanish] provinces have hitherto been carried on . . . on the second of January last a Company of Spaniards, whilst travelling on the highway leading from the Sabine to Bayou Pierre, were attacked by a party of Said Banditti about thirty in number, with their faces blacked, and otherwise disfigured, who fired upon them, killed one man, wounded several, one it is supposed mortally so, and robbed them of all their horses, mules, baggage and about six thousand dollars in specie.[43]

With the amount of trade along the road and the lawlessness in the Neutral Ground, the passage between Spain and the United States through Nacogdoches was a treacherous and costly one.

Events beyond Control
1811—New Madrid Earthquake

In the midst of boundary disputes, immigration, and growing illegal trade into temptingly accessible territory, an unpredictable natural event added to the mix of pressures on Spain and led to increased southerly migration from Missouri and Arkansas toward Spanish Texas. During the final months of 1811 and all the way through March 1812, a series of high-magnitude earthquakes devastated land in southern Missouri. The quakes created lakes where there were none before and caused rivers to run backward for a time. The town of New Madrid, Missouri, was the epicenter, but many other small settlements within a 250-mile radius were completely destroyed.[44] Families on the White River where Trammell settled were shaken from their homes. One of those families was the Laffertys. They migrated previously to the Missouri Territory with Nicholas Trammell and were associated with him in later accusations of wrongdoing. The Laffertys felt the full force of the tremor. "On South Bank of the White River old man Lafferty settled opposite the south mouth of Lafferty Creek, bringing with him from East Mississippi River, considerable livestock. He was living at this place in the time of the earth quake of 1811, and reported he saw the bottomless well near his landing blown out. He said there came a terrific shock, and muddy water raised from one side of the river to the other with a great explosion near the south bank. . . ."[45]

The resulting southerly movement of people out of the devastated landscape further opened the Southwest Trail from St. Louis as far as the Ouachita River across the middle part of Arkansas. Trammell and his kin were settled all along the White River in northeastern Arkansas and were no doubt affected by the earthquakes and the resultant urge to move out of the region. There is no record of Trammell's immediate relocation, but the movement of people southward and his resultant development of the trail that became Trammel's Trace closely tie the two events chronologically. Another element of the disaster created insecurity beyond the physical destruction. The Cherokee chief, Tecumseh, asked the Great Spirit to bring destruction on invading whites. Some believed the quake was the Great

Spirit's signal to all the tribes to unite against the whites who had stolen their land.

A further development in the hot pursuit of property in the emerging territory also came as a result of the earthquake. To compensate landholders in the area impacted by the earthquake, Congress issued 515 New Madrid land certificates. The "floating" certificates entitled the bearer to locate unclaimed public land virtually anywhere and lay claim to it. Instead of helping those who lost land, the certificates opened doors to fraud and speculation. Before the victims had time to learn about the passage of the legislation intended for their benefit, the Missouri Territory filled with opportunists. Only twenty New Madrid certificates were used by the original grantees. The rest were sold by the bearers to land speculators. The land where Little Rock, Hot Springs, and Fulton are now located was claimed using purchased New Madrid certificates.[46]

As the trail toward southwestern Arkansas to the edge of Texas opened to more trade and traffic, Nicholas Trammell certainly would have observed that movement, even if he was not a part of it. With the events of 1811 pushing others southward and the lure of opportunity drawing him in from the direction of Spanish Texas, the impetus for Trammell's migration farther south was firmly in place.

1811—Pecan Point

The eastern borderland with the United States was not the only pressure point in the growing diplomatic and geopolitical difficulties on the edges of Spanish Texas. Immigrants and squatters from the growing republic to the north tiptoed across the Red River and settled on the south bank in northeastern Texas, on the Spanish side of that yet-to-be-determined boundary. John Sibley described an Anglo settlement that began to form in 1811 at a place called Pecan Point.

> I understand there Number is about Twenty and are daily Augmenting, Some of them from this place [Louisiana], Some from Arkensa, Washita & Most of them have Escaped from different Jails in the United States, the Indians have repeatedly Complained to me of their Ill Conduct, they are enemies to all Law and Good Order, and If suffered to remain there long Undisturbed they will become so Strong that it will Cost the Government an expensive expedition to remove them. They are building Cabbins and Planting Corn & doing Great Mischief Among the Indians,

> And Inticing Negroes from their Masters & enticing them as comrades. The Indians in this quarter are all quiet & friendly to our Government & would generally Conduct well was it not for their having too much Intercourse with Bad white People, with which this frontier of the United States is too much Invested.[47]

Whether Sibley's report on the criminal pasts of the settlers was completely factual or an exaggeration of the unknown is uncertain. What is clear, however, is that this reputation of the Pecan Point settlers lingered for years.

Geographically, settlement was widely dispersed along the Red River. Pecan Point was a peninsula formed by a loop of the Red River covered by pecan groves and a well-known buffalo crossing. Its first documented mention is in the French archives at Natchitoches, Louisiana, where a campsite called "Pointe aux Peconques" was noted on the route to the upper Red River, at the same location as a former home for a branch of Caddo Indians. Six months after his earlier observation, Sibley again detailed the dangers of the Anglos along the Red River.

> There is a party of Bad Man fugitives from the different jails in the United States who have Settled themselves at the Pecan Point on Red River about 500 Miles by Water from this Town. The Names of those I have heard of, are Glover, Colton, Parkham [Barkham or Barkman], Armstrong, Coots, Harper, Gibbons, Kelly, Fouts, Turner, Rogers, Patton, Lucas, Williams, Dixon, Knowlton, Spears, and some Spaniards & Runaway Negroes, these people are Enemies to all law and good order, and most of them would have been hanged If they had have remained in the United States. The Indians have Killed one of them by the Name of Dixon. They have imposed upon the Indians by Forged passports, they are all Murderers, Thieves & Robbers, and doing all the mischief they Can Amongst the Indians, their party is Augmenting and will give us Trouble before long, if they are not broken up. They are planting Corn &c & appear as though they Intended to Settle themselves Permanently, which will be an Assylum for Runaway negroes & all Bad people.[48]

The earthquake shook up more than the land around New Madrid, Missouri. Although Nicholas Trammell was not mentioned as one of the earliest bad men along the Red River, some of the names Sibley noted were from the same areas of Kentucky, Tennessee, and Missouri where Trammell lived and traded. Reports of trading activity from the earliest Pecan Point settlers would certainly have filtered back to Trammell in the Missouri Territory. The numbers of mustangs in the region and the natural shoal where buffalo in great numbers crossed the Red River two times a year were attractive to traders and hunters. Although Trammell appears to have remained at the White River for the time being, his connections and relations to people who were moving to Pecan Point offered him new opportunities for trade and a new vantage point on the edge of Spanish Texas.

That Sibley voiced his concerns about so few settlers 500 miles from Natchitoches in fairly serious tones was a testament to his understanding of the dynamics of the borderland. Every encroachment on territory under dispute or clearly in Spanish hands emboldened Americans to take over even more territory to the west of the Louisiana Purchase. From the perspective of the United States, the settlement at Pecan Point was a tear in the fabric of the law. It was a place where money and perhaps lives would have to be invested to maintain the peace. From the Spanish point of view, it was simply further evidence of the land-hungry power of the United States.

1812—Gutiérrez-Magee

Political turmoil from within and increasing pressure along its boundary left Spanish Texas vulnerable. Before 1812, the authorities were content to confiscate the goods of illegal traders and jail smugglers. In April 1812, they became more aggressive in response to smuggling and unlawful settlements along their boundary with the United States. A patrol of twenty soldiers marched from Nacogdoches with the assignment to burn any unauthorized dwellings and stop all cultivation between Nacogdoches and the Sabine River.

In spite of the posturing, Spain's presence in East Texas was weakening. Rumors persisted for years about the plans of armed invaders from the United States to enter Spain and capture territory. It was not necessarily an invasion by the US military that was the threat; rather, it was a takeover by land-hungry men with a thirst for power that threatened the region around Nacogdoches. That imbalance of power created an environment that led to the Gutiérrez-Magee Rebellion in August 1812.

Don José Bernardo Maximiliano Gutiérrez de Lara was a passionate proponent for Mexican independence from Spain who was forced eastward from San Antonio as a result of his dissent. Augustus Magee was a first lieutenant of artillery in the US Army stationed at Natchitoches. Gutiérrez met Magee in Natchitoches, where they planned their invasion and openly advertised for recruits. As part of Magee's military duties, he had arrested twelve to fifteen criminals who were operating in the Neutral Ground on the east side of the Sabine. Rather than imprison the skilled and resourceful outlaws, Magee enlisted them as mercenaries. William Barr's trading partner with the Spanish, Samuel Davenport, also proved his alliances were largely of convenience. Davenport joined the Magee rebellion against his Spanish trading partners, which resulted in Spain issuing a reward for his head.[49]

One measure of the Anglo perception of Spanish weakness at the border was the size of the force Gutiérrez and Magee thought was needed to gain control over the Spanish settlements in eastern Texas. They invaded the Spanish garrison at Nacogdoches and proclaimed Texas free from Spain on August 12, 1812, with only 150 men. The Old Stone Fort served as their headquarters.[50] They took Nacogdoches without resistance even though their movement across the border from the United States was no secret. They sought support quite publicly, and some in Texas welcomed them as heroes.

Spain was in retreat. Nacogdoches was abandoned and devoid of Spanish troops. Trinidad was abandoned and burned to the ground the following year by Spanish troops.[51] This set the stage for the emergence of Trammel's Trace to the north as an alternate route for smuggling. With the Neutral Ground written off to lawlessness and Nacogdoches emptied of the remains of Spanish influence, what remained was an open door to trade through Nacogdoches across the border to Natchitoches—a door at the end of a secretive trail that became known as Trammel's Trace. By 1813, many factors came together that encouraged Nicholas Trammell to cut Trammel's Trace along that unpatrolled no man's land, smuggling horses and goods from the north to Nacogdoches and along the El Camino Real. There were no settlements to reveal his activities, no soldiers to monitor his movements or confiscate his property, no laws governing his enterprise, nothing other than his own ambition and his willingness to take a calculated risk.

5

1813–1819
Couriers of the Forest

> ... *this trade is a species of land piracy which is carried on by those traders against the citizens of this nation, for it amounts to the same thing whether those traders plunder mules themselves or hire the Indians to do it.*
>
> —STEPHEN F. AUSTIN[1]

Travel between Louisiana and Texas in the early 1800s was travel between two nations, Spain and the United States. Matters of trade, land claims, settlement, and travel were international issues in the region that is now Arkansas, Louisiana, and Texas. What seemed simple commercial trade just across the Sabine in the United States was illegal smuggling between nations. People involved in illicit trade like Nicholas Trammell were called freebooters—the name given to anyone who lived by taking advantage of others. Lone bandits, former privateers, groups bound by criminal enterprise, and adventurers who seemed to have some sanction by the US government roamed across the ill-defined borders between the United States and Spanish Texas, in and out of disputed lands. Freebooters were simply part of the criminal and commercial enterprise of the day. Smugglers did not necessarily consider themselves outlaws, even when they intentionally avoided the authorities. They were, in their own eyes, merchants who took risks for profits. They were *coureurs de bois*—couriers of the forest.[2]

Avoiding detection was part of the history of commerce in the area for at least a hundred years before Trammel's Trace connected the Southwest Trail with the El Camino Real at Nacogdoches. One hundred years before

Trammel's Trace was used by US traders, Louis Juchereau de St. Denis obtained consent from French authorities in 1714 to go from Natchitoches into New Spain (what would become Texas) to open trade relations between the French and Spanish colonies. The provincial governor pronounced St. Denis's effort a piece of insanity. After St. Denis left Natchitoches and crossed the Attoyac River, he continued due west rather than turning northwest to Nacogdoches. This route allowed him to move his caravan of merchandise south of Nacogdoches, avoiding the Spanish authorities located there and the payment of duties on his trade goods. St. Denis's road bypassing Nacogdoches was used for moving goods around the authorities from 1690 until well after the beginning of the nineteenth century. Using a way around was a common tactic for moving goods, and smugglers were always interested in alternate routes.[3]

Roads and the Westward Expansion

Early roads that began for a single purpose could quickly become more formalized and established. An example of this was the Natchez Trace, a well-known and heavily used road that connected southern stretches of the Mississippi River at Natchez to salt licks near Nashville, Tennessee. The Natchez Trace experienced its heaviest use from 1785 to 1820, much of it by the Kaintuck (Kentucky) boatmen who floated down the Ohio and Mississippi Rivers to markets in Natchez and New Orleans. When they arrived in Natchez, having no power to move their flatboats back upriver, they sold their cargo, scrapped their boats for lumber, and began the 450-mile trek back to Nashville on foot. Despite its frequent use, the Natchez Trace did not come by the name we know today until well after the wave of traffic began to ebb around 1826. Earlier accounts refer to it as the Path to the Choctaw Nation, the Choctaw-Chickasaw Trail, or the Chickasaw Trace. Nearby residents knew it as the Natchez Road or the Nashville Road, depending on whether one's travel was north or south.

Traders and merchants along the Natchez Trace provided creature comforts, called stands, in the form of taverns, hotels, and supply depots for the hundreds of travelers who were potential customers. The volume of trade led to a trade route that was well developed and maintained to cater to the large number of routine travelers.

The undeveloped, backwoods cousin of the Natchez Trace to the west had no such improvements in its earliest years. Scattered Indian villages where traders might exchange blankets, cooking utensils, or weapons in trade for horses or hides were the extent of the marketplace in the new frontier. Other than Nacogdoches, there were no other Anglo settlements along Trammel's Trace in the earliest years of its use. Nacogdoches was virtually abandoned between 1812 and 1820 because of attempts by filibusters to invade the Spanish borderlands, so even the terminus of Nacogdoches offered little in the way of supplies.

Unlike the Natchez Trace, which was a road for legitimate commercial enterprise, Trammel's Trace initially served as a primary route only for those who wanted to avoid the entanglement of law. By 1817, Tennesseans improved the Natchez Trace by erecting mileposts along the way, showing distances to courthouses or the nearest town. Trammel's Trace in the same year was still only wide enough to accommodate people on horseback, not wagons or teams of horses to pull them. By 1820, the hundreds of people traveling the Natchez Trace became thousands, and wagon traffic wore the trail down to the point where it sank twenty feet below ground level in places.[4]

At some point between about 1813 and 1815, Nicholas Trammell began traveling farther south from northeastern Arkansas into eastern Texas for his business ventures.[5] Exactly when he started using the old Indian trails cannot be certain, but the nature of his business was about to be exposed to the governor of the Missouri Territory.

On the Record as Horse Thief

Stories about the thousands of mustangs of Spanish and Arabian descent that roamed the prairies of Spanish Texas stirred the imagination of a nation that moved by horses. The relationship between a man of the frontier and his horse was more than a relationship of necessity. In the early part of the nineteenth century, a man's horse was everything—his friend, his protector, even his liberator when he needed speed for an escape. Riding alone in unexplored territory where Indian troubles could arise at a moment's notice required a horse that could outrace an adversary. When traders and adventurers made trips of hundreds of miles into unknown territory, a horse that could endure the trip with little more than the grass and water available along the way was a necessity. Civilization may follow the plow, but in the settlement of southwest Arkansas and northeast Texas, the plow horse came long after the mounts of others left their hoof prints across the prairies and piney woods. Before settlers braved the new territory, long hunters, mustangers, and freebooters rode for years in search of game and trade through the inhospitable wilds crossed by Trammel's Trace.

The importance of hoofed labor to every part of life meant that horses and mules were highly valued. Trade on the frontier was no different than trade today—a transaction that required two parties. Commerce is dependent on customers. In the early 1800s, that commerce frequently involved trade in horses that Indians stole from Anglos and competing tribes, with men like Trammell smuggling them into the United States, where they were sold for trade goods to restart the business cycle. Nicholas Trammell Jr. got his start in Arkansas as a trader with more general business activities than horses, but his interests later turned toward the simpler method of expecting his Indian partners to do the dirty work of stealing horses.[6] Trammell's business formula was ideal for the unsettled times. He found favor with the Cherokees in Arkansas by providing them with cloth, cooking utensils, and perhaps even weapons in exchange for furs and horses. In return, he gained a profitable commodity with little risk.

All tribes, even the more civilized agrarians, relied on horse thievery. Some tribes engaged in the theft for sport, returning the horses grudgingly if they were discovered. One particularly successful technique was to steal horses that had been on the trail all day while they were tired and hobbled for the night. When horses disappeared, a search party formed immediately to seek the return of property or the blood of the culprits. One diarist's account told of his encounter with hunters who were up by daylight to recover horses stolen from them by Cherokees the night before.

Trammell and his allies took advantage of the Indian talent for horse thievery and directed any blame toward the Indians. By accepting stolen horses in trade, Trammell could feign any involvement in the theft itself. Perhaps that plan between trading partners worked for a time, but it is likely that neither Trammell nor his Cherokee partners could resist an opportunity to gain an advantage. When the parties in business together were both engaged in illegal activity, gentlemen's agreements did not always last. Those frontier pacts often became quite troublesome, as evidenced by Trammell's first mention in connection with the activity for which he was best known. On April 27, 1813, Trammell's Cherokee neighbors near the White River accused him and his half brother, Mote Askey, of more than just stealing a few horses.

> Address of Cherokees
> To his Excellency Benjamin Howard Governor in & over the Territory of Missouri; Father we are a part of the Cherokees tribe of Indians; have settled on the White River a water of the Mississippi by we presume,

the consent of the Government of the United States, where we are endeavouring to cultivate the soil for our support & wish to live uninterrupted by the malicious white people; but the revurse; there are a few bad men combined together for the purpose of stealing our horses & do steal them; to wit; Nicholas Trammel Mote Askey John Wells Joseph Carnes Robert Armstrong. Barnet Peter Tileo Thoms James John D. Chisolm Dennis Chisholm Ignatious Chisholm Jessey Isaacs; John Williams Robert Trimble William Trimble; William Smith John Lafferty Senr. & Ace Musick, ____ Pain & Joseph Pain; all of which are on publick lands; we pray they may be removed from amongst us; if it be consistent with Government; some of which characters have solicited us to join them in killing robbing & burning the Houses of the honest & industrious part of the white inhabitance neare to us; which we wish to live peaceable with; to wit, Nicholas Trammel & Mote Askey; we have lost by those characters Twenty Horses in course of Twelve months & if some measures are not taken we shall in a short time be left destitute of property. . . ."[7]

This terse and pointed accusation followed an earlier warning. One of the Cherokee signatories to the petition, Connetoo (aka John Hill), previously advised Indian agent Return Meigs about Anglo thefts: "You well Know that when there is a band of bad men it requires exertion to supporess them. It appears they intend to lead me into distress. Likewise they are corrupting my own people. It is a very disagreeable thing to the Chiefs that their young men should be curroupted by the whites."[8]

What seemed like blunt, legal assertions by the Cherokee were in fact telling examples of the blurred lines between what was legal and illegal and the tenuous intermingling of the Anglo and Indian residents of that region. Even though their relationships were generally peaceful, they were not without conflict.

The primary Cherokee accuser was Chief Connetoo. Connetoo was an ambitious chief of mixed blood, as well as an agent for US trading companies with his own economic interests to protect. His control of an immense amount of fur trade put him in direct conflict with Trammell and other unlicensed traders working for a piece of the action.[9] Connetoo worked diligently to portray an image of doing the right thing even while the Anglos were not. In March 1813, Connetoo wrote Indian agent Meigs to inform him that his Cherokees were weaving cloth as a way to extend their civilized nature and avoid the troubles inflicted by unscrupulous Anglos. In the same month, Connetoo pointed out the ongoing abuses by Americans in simple terms: "When a white Man enters my house I treat him in a stile to render him comfort. When he leaves my house he fails not to run off with some of my property."[10]

In the midst of this dispute, William L. Lovely was newly assigned as agent to the Arkansas Cherokee in July 1813. He quickly recognized that there were also broadening conflicts between the Osage and Cherokee. Tensions between these two tribes increased when the federal government moved the Cherokee to land in Arkansas that was already the hunting grounds of the Osage. Lovely also recognized that this was not simply a problem with the Indians. He reported that white settlers living in the area were behaving in a way that was dangerous to the peace. A later account from a Cherokee chief hinted at the methods of deception used to entrap the Indians: "There are some white men who sit near the edges of our country, who steal our horses, cattle and hogs, who lay whiskey there. These rob us and impoverish us."[11]

Many of the others accused by the Cherokees along with Nicholas Trammell and Mote Askey were documented residents of the White River settlements, some of them prominent families. Wells, Williams, Trimble, Musick, and Smith are listed along with Nicholas Trammell on the 1815 tax list for what was then Lawrence County, Missouri. John Williams became a constable and justice of the peace in the 1820s. Asa Musick and his brother, Robert, were traders who moved with the commercial opportunity. Asa stayed in northern Arkansas, but his brother Robert moved with the migration. Asa later opened a trading factory in southwest Arkansas not far off the roads regularly used by Nicholas Trammell.

John Lafferty was one of many settlers from Tennessee and Kentucky who came to the White River at the same time as Nicholas Trammell. Lafferty was a trapper and hunter actively engaged in trade with the Indians along the navigable waterways. Like Trammell, Lafferty laid claim to 640 acres of White River bottomland in 1810. He was identified by a contemporary as a friend of the Cherokees who provided them with a fair trade, making it all the more curious that he was faced with such allegations by members of the same tribe.[12]

The Trimbles from Livingston County, Kentucky, were also part of the large group of families who migrated with Lafferty, first by wagon and then by boat upriver from the Arkansas Post. They herded the group's cattle cross-country while others traveled by boat to the new settlement at the mouth of Poke Bayou on the White River.[13] William Trimble was a prominent attorney who later served in the Arkansas Territorial Legislature in

1823 and again in 1831. He also became a judge in the Arkansas Superior Court in the 1830s and acted in that role in an 1831 case involving none other than Nicholas Trammell.[14]

The most intriguing of the characters accused by the Cherokees were John D. Chisholm and his sons, Dennis and Ignatious. John Chisholm was a noted Cherokee chief who led the first band of Cherokees from Alabama and Tennessee to settle on the White River. Another son of John D. Chisholm not mentioned in the accusation, Thomas Chisholm, was presented with a medal by Thomas Jefferson in 1809. John later represented the tribe when a treaty was negotiated with the United States in 1817. Being a trader in the frontier often meant building roads to improve trade routes, a trait the Chisholms shared with the Trammells. John's grandson, Jesse, was associated with the marking of the famous Chisholm Trail in 1864.

Indian agent William Lovely had little respect for Chisholm in spite of his prominence and his representation of the tribe to the United States. Lovely's handwritten note on a letter from Chisholm to Agent Meigs referred to Chisholm as a "puppy [puppet] and a raskal."[15] Before Chisholm was accused of horse thievery with the Anglos, he was sued in the courts of the Arkansas Territory for nonpayment of a trading debt accumulated between March and September 1810. Interestingly, his purchases included not only beef and pork and some finer goods like calico and muslin but also significant quantities of whiskey. Perhaps John Chisholm used the whiskey in just the ways the Cherokee petitioners described.

Although the list of accused on the Cherokee petition to the governor included both Anglos and Cherokees who were prominent citizens, the person with the most credibility and integrity was on the list of thirty-three Cherokees who signed the petition against Trammell and Askey. On that list was Thomas Graves, a noted Cherokee chief honored by President Jefferson. Graves was one of the elite Cherokee chiefs, a reputable and cooperative leader of an agriculturally oriented people. The Cherokees in the Missouri Territory were civilized to an extent that contrasted with the ways of the poor, white settlers. The stereotype of savage Indians and civilized whites was not the case when it came to the Cherokees under Graves's leadership.[16]

When Graves and the other Cherokees put their complaint about Trammell in writing, they provided Governor Howard with a calculated incentive to act. Accusing Trammell and Askey of stealing horses alone probably surprised no one. That frontier crime was widely ignored and selectively enforced. They specifically mentioned Trammell and his half

```
1810            John D. Chisholm            Dr
March
  6    To 2 1/2 Gallons Whiskey ----------  5 " 00
  9    To 104 lb porke ----------------    7 " 28
 10    to  5 quarts Whisky ---------        2 " 50
 25    to 3 Gallons Whiskey at Sundries times  6 " 00
 27    One sow and pigs ------------       10 " 00
 27    - 90 lb porke --------------         6 " 30
April
 19th   76 lb Beef ------------            3 " 80
        1 1/2 Gallons Whiskey --------     3 " 00
 27th May 50 lb Beef --------              2 " 50
        50 lb salt  had some               5 " 00
        3 quarts     Whisky -----          1 " 50
        3 1/2 yds Callico -----            7 " 00
        3 " Do Muslin -----                3 " 00
June 11th 71 lb Beef                       3 " 35
        42 " Do ----                       2 " 10
        2 quarts whisky for By Hall --     1 " 00
        2 1/2 quarts Do                    1 " 25
7th July 154 lb Bacon - fat Bacon -       19 " 25
        1 Drest Deerskin --------          2 " 00
        2 Hats ---------                  10 " 00
        Bayd for Wells ------             20 " 00
        1 quart Whiskey                     " 50
        16 Bushels Corn --                16 " 00
        part of Blanket ---                3 " 00
Sept   2 fat Hogs -----                   10 " 00
        6 lb powder ----                    6 " 00
        pdt of Heone                       20 " 00
        1 Boat                             30 " 00
        For work Which was not performed  25 " 00
        To 80 Bushels of Corn @ $ 50/¢ Bushel 40
        Errors Excepted        Robert           07 " 33
```

Liquor, supplies in trade, 1810. John D. Chisholm.

brother Mote Askey a second time to call attention to a more serious concern that guaranteed Governor Howard's attention. The Cherokees said that Trammell and Askey "solicited us to join them in killing robbing & burning the Houses of the honest & industrious part of the white inhabitance neare to us."[17] The petition was not just about losing twenty horses in a disagreement over trade; it was about Trammell and Askey advocating violence in the region, actions that no doubt unsettled the governor. Although Lafferty, Trimble, Chisholm, and the others were not directly accused of the more serious crimes, the complaint made clear the Cherokee's belief in their guilt through association. Trammell and Askey were the certifiable outlaws, and perhaps managing partners, of the horse thievery and trade that included all the others. By pointing out that Anglos were really to blame for inciting crimes more commonly assigned to the "savages," the crafty Chief Graves hinted that the Cherokees might join the side of bloodshed if the government did not respect them and deal with Trammell. William L. Lovely weighed in with a view of the conflicts that did not look with favor toward the Anglos: "I am here without a cent, and among the worst banditi; all the white folks, a few excepted, have made their escape to this Country guilty of the most horrid crimes and are now depredating on the Osages & other tribes, taking off 30 horses at a time, which will show the necessity of giving some protection to this place...."[18]

This conflict continued into 1814. Lovely suggested that a military presence could help stop further battles between the Osage and Cherokee and put a stop to the illegal trade and horse thievery. Lovely asked for two companies of troops to come to the region, not due to threats from the Indians but to control "some white of the worst character in this country whose influence with the Indians is dangerous to the peace of the same."[19] Three years after his request, the government constructed Fort Smith in western Arkansas to develop that military presence. There is no record of any action taken in response to the Cherokee's complaint to the governor about Trammell and Askey.

In the meantime, the smuggling of horses all across the region continued. It was not only Anglos and Indians who traded horses and mules across the porous borders with the United States. Mexicans who struggled for independence from Spain also engaged in smuggling largely out of economic necessity. The Spanish government tried to stop the trade within Texas by declaring all wild livestock property of the Crown. Nevertheless, prominent families in San Antonio banded together to take horses and mules to Louisiana in trade for needed goods. José Antonio Navarro, a

Tejano who later became one of the signers of the Texas Declaration of Independence, was one of those involved.

Navarro lived in New Orleans for several years and gained a sense of the commercial opportunities such trade ventures allowed. He returned home to San Antonio in 1816 after his mother secured a pardon for his rebelliousness against the ruling authorities.[20] The family's home was ruined, and they were in dire financial straits. Navarro resorted to what many others did to survive: he smuggled mustangs and mules across the border to Louisiana from La Bahia (Goliad) using the Contraband Trail that ran from San Antonio, through Gonzales, and then met up with the middle crossing of the Trinity River on the El Camino Real. Navarro was captured and jailed by Spanish authorities in 1819 for "going into the interior without a license," an offense directed at preventing mustanging ventures. From his perspective, José Antonio Navarro only tried to do what many Tejanos were trying to do—provide for his family.[21]

The Resettlement of Punta Pecana

Pecan Point was a geographic magnet for early traders. The mouth of Pecan Bayou at the Red River created a shoal—a natural crossing for buffalo, mustangs, and the men who pursued them. The attraction of the early Anglo traders to this particular place was strong in spite of its remoteness and distance from any other settlements and perhaps also because of the political murkiness between nations. Whether people along the Red River were Anglo or Indian, they were there for mustangs, buffalo, and trading partners. The unlicensed traders had a huge advantage over posts much farther away. By conducting fur trade at Pecan Point near the hunting grounds, only five traders captured the market for all of the Indian trade of that region. The sanctioned Natchitoches traders far to the east realized they were losing Indian business to these unlicensed traders. If trade was unsettled, the population in general was as well, both Indian and Anglo. A bigger problem for the Anglo settlers at Pecan Point was their settlement in a borderland in dispute between Spain and the United States and in an area hunted by more than one Indian tribe. Anglo settlers were forced out as a consequence of pressure from both the United States and Spain.

The desirability of the crossing at Pecan Point on the Red River brought them back, attracting others to return and resettle there. After the war with Great Britain ended in 1815 (War of 1812), there was a fresh movement of settlers up the Red River, both overland and upstream along the difficult, shallow waterways.²² George and Alex Wetmore were former sutlers with the US Army during that war. In June 1815 they established a trading post at Pecan Point. The Wetmores were followed by William Mabbitt, formerly of the Arkansas Post, who set up his own trading house only a few months later. The following year, more permanent Anglo settlement began with the arrival of Walter Pool, Charles Burkham, and Claiborne Wright.

In 1818 only five Indian traders and twelve Anglo families lived at Pecan Point.²³ Traders at Pecan Point hoped to preempt some of the trade going to Natchitoches by placing themselves closer to the Indian country to the west, even if they defied laws of the United States and Spain in order to do so. When traders made their settlement near the hunting grounds of the Caddo Indians at Pecan Point, a Caddo chief complained that the Anglos occupied "the only crossing place for Buffalo for many miles, and the only crossing from which his people derive any advantage from their crossing."²⁴ The Osage were more aggressive in their response to the encroachment. In the summer and fall of 1815, the Osage stole sixty horses from whites along the Red River, and Anglo hunting parties were attacked and violently killed.

Whether they were on the north bank of the Red River or the south, Anglo settlers were in disputed territory. If they were on the south side, they were in Spanish Texas, where they had no legal standing. If on the north, they were in the public lands of the United States designated for the Osage and prohibited from being there by the terms of Indian trade laws established by Congress. Either way, they were at the farthest reaches of the United States, and it was only that geographic reality that prevented even more trouble than what was to come.

Objective observations of these lingering conflicts were difficult to obtain, but one account offered some insight. Thomas Nuttall was a British botanist who accompanied US soldiers in their eviction of Red River settlers along the Kiamichi River and Pecan Bayou. Nuttall offered his observation that it was the policy of the United States that contributed to the disturbed state of affairs on the frontier. With Indians pushed farther to the west in resettlement, he acknowledged the "universal complaint of showing unnecessary and ill-timed favours to the Indians" over the Anglos who had settled there without legal authority. The possibility of military intervention did little to counter the Indians' ability to commit robberies, murders, and other depredations on the Red River settlers. Nuttall's observations of the impact of this policy on the evicted Anglo families whose homes were burned by US soldiers were direct and prescient: "It is now also the intention of the United States government, to bring together, as much as possible, the savages beyond the frontier, and thus to render them, in all probability, belligerent to each other, and to the civilized settlements which they border. To strengthen the hands of an enemy by conceding to them positions favourable to their designs, must certainly be far removed from prudence and good policy."[25]

Nuttall noted that to have left the Indians in their ancient eastern homes would have surrounded them with civilization, forcing them to assimilate or at least to restrain their aggression. In that position, there would be no need for an expensive military agency focused on controlling them or for frontier forts for the purpose of "coercing them by terror."[26]

In spite of the conflict, residents of Lawrence County in the Missouri Territory along the White River began to move to Pecan Point by 1815, possibly as a result of Nicholas Trammell's trading ventures to the Red River and beyond there to Chihuahua, Mexico. The connections of the Pecan Point settlers to Nicholas Trammell went back to Logan County, Kentucky. Adam Lawrence was one of the earliest settlers who moved even farther up the Red River from Pecan Point by 1815 to the place where Jonesborough formed. Daniel Davis settled his family at Jonesborough in 1818. Davis later traveled down Trammel's Trace from Pecan Point with his belongings and livestock to the Ayish Bayou district near the Sabine, east of Nacogdoches.[27]

Claiborne Wright arrived at Pecan Point on September 6, 1816, after a difficult journey of six months by boat.[28] Wright previously worked with Trammell's uncle Philip at his salt works in Tennessee. Wright's trip to Pecan Point was an arduous one. From Tennessee, Wright and his large party easily floated downstream with the current on the Ohio River to the Mississippi, a trip of about a week. Past Natchez and down to the mouth of the Red River, large plantations developed and the banks of the river showed signs of settlements and farms. At that point, however, the journey took on a completely different character. To get where they were going, Wright and his family had to travel upriver against the current.

The vessel of choice for this kind of river travel was called a keelboat. Keelboats were amazingly versatile vessels, designed to make the best of their ability to carry heavy cargo and passengers through shallow water. Keelboats had sails, when that form of power was available, and a mast with a hinge that allowed it to be lowered when needed. When the wind was not cooperative, the keelboats were rowed, poled, or towed upstream by the use of a rawhide towline called a cordelle. The cordelle was fastened to the boat and pulled by as many as twenty men, or women, as needed, walking along the riverbank. More often the keelboats were poled upstream against the current. Men passed from bow to stern with the pole firmly dug into the bottom, walking on a narrow running board. Keeping the bow into the current was critical, for if it swerved off center, it was difficult to control. By staying close to the bank, the keelboat could move from one eddy to the next. However, that was also where the snags and stumps were, and they made travel more difficult. If the keelboat had to cross the current to gain some headway, it could lose a quarter mile of progress to the downstream flow. Forward progress of no more than one mile per hour was expected.[29]

William Dewees described the difficulties of his journey up the Red River on a keelboat: "After entering Red river, we found our labors very toilsome; on account of our boat being a large, family boat, crowded with women and children, we found it very difficult to row and push upstream. However, we got along very well, though slowly, until we arrived at the Big Raft."[30]

The Great Raft was a logjam of historic proportions on the Red River. Snags and trees caught debris as the logjam moved downstream. Floodwater ripped entire trees from the sandy banks and deposited them in an ever-growing mass of logs that blocked the natural flow, creating many side

The Great Raft in the Red River. Photo by R. B. Taylor, courtesy Northwest Louisiana Archives at LSU Shreveport.

streams and swamps. Historians estimate the logjam began to form around 1100–1200 AD. Its lower end was about ten miles upstream from Natchitoches. By 1806, it stretched one hundred miles upriver. Dewee's group took thirty days to go only ninety miles through the Great Raft. Even the hardy and well-manned expedition by Freeman and Custis in 1806 described the incessant toil, fatigue, and uncertainty it took to move through the Great Raft and the Great Swamp that surrounded it.

Indian agent Sibley estimated that four hundred families were scattered in the land along the Red River reaching up to Pecan Point by 1817 and expected the numbers would double in one year.[31] The large numbers of families who moved into the disputed boundary region were an increasing concern for the governments of both the United States and Spain. The United States had already promised land along Red River, convincing Indian groups from the eastern states to move west. When the tribes relocated and found Anglo squatters and traders already on the land, the whole region was ill at ease. In response to the violation of land agreements and illicit trade on the Red River, US Army Major Riddle set out for Pecan Point in July 1817 with fifty soldiers from the 8th Infantry stationed at Natchitoches. His orders were to put a stop to the illegal trade and arrest those engaged in that business.[32]

Pecan Point was hundreds of miles from Natchitoches. One possible route was the network of roads that led up Trammel's Trace from Nacogdoches and then along part of the old road to the Spanish Bluff east of Pecan Point. The road was formed when Spanish soldiers were sent to intercept Freeman and Custis. However, US soldiers could not use that road since it was in Spanish territory. Instead, they were required to remain on the US side of the Red River, making it likely Riddle traveled up the east side of the Red River and swung west around the Great Bend through what is now southeastern Oklahoma.

During his trip, Major Riddle arrested Francis Dursey and John Petty, confederates of the trader William Mabbitt, and confiscated about $2,000 worth of goods. Dursey escaped, but on the way back to Natchitoches Riddle captured Mabbitt himself, probably on the lower reaches of the Red River. Mabbitt and his crew were jailed in New Orleans, and his goods were confiscated until their status could be adjudicated.[33] Although Riddle's arrest made an impression, the military effort had little lasting impact. Pecan Point was simply too distant from US authority in Natchitoches. That distance was a major contributing factor in the ongoing failure of attempts to close down trade and end the settlements around Pecan Point.

When arrests did not interrupt trade, the US military escalated its efforts and began to physically evict US citizens from land they had settled and cultivated for years along the Red River at Pecan Point. The United States granted the land along the Red River to Choctaw Indians in order to evacuate them from Mississippi. The Anglo evictions secured the land promised to eastern Indians who agreed to resettle in the West. The Choctaw were given vast acreage by the US government, and that same land was claimed by the Osage for hunting grounds. When Anglos settled in the same territory, it only served to heighten tensions.

Settlement continued unabated, and in May 1819, Maj. William Bradford proceeded from Fort Smith in Arkansas to evict the families living on the south side of the Red River. The soldiers crossed tick-infested buffalo traces and unmarked thickets to arrive at the home of William Styles. Styles's wagon left a trail, and the soldiers found him living on the edge of the prairie, along with his wife and his blind, ninety-year-old mother-in-law. Styles was only about five miles from the mouth of the Kiamichi, where it emptied into the Red River. The soldiers informed Styles he would have to leave the land. Fifteen miles upriver, the troops gave the same bad news to Martin Varner. Varner lived on the edge of Horse Prairie, so named for the herds of wild horses that attracted the attention of traders from all nations. A member of the soldiers' party noted, "These people, as well as the generality of those who, till lately, inhabited the banks of the Arkansa, bear the worst moral character imaginable, being many of them renegadoes from justice, and such as have forfeited the esteem of civilized society."[34] At first the military simply explained the problem and allowed the settlers time to bring in their crops. With that approach, the military's relationship with the settlers could be amiable. The congeniality changed, however, when Major Bradford returned to burn houses on both sides of the Red River, leaving settlers in the Indian land west of the Kiamichi without homes or protection. The Pecan Point settlers who had cleared land for farming were unexpectedly thrust out into the wilderness.[35]

Horse Thief, Gambler, and Methodist?

Although the reputation of the Pecan Point settlers was that of bad men and outlaws who were there only to escape legal problems and debts in the United States, they were not without enticements for behavior of a different sort. Religion followed them to the Red River. William Stevenson, a former Tennessean and Methodist minister, moved to southwestern Arkansas in 1814 to join his brother. Upon his arrival he began preaching and by the following year was ministering to his fellow Tennesseans at Pecan Point. Claiborne Wright's move to Pecan Point the following year may have been a result of communication from Stevenson back to Smith County, Tennessee. Wright was from the same county in Tennessee as Stevenson. When the Wrights settled in Pecan Point, their home became a place for the Methodist minister to preach and evangelize.[36]

Methodism followed the settlers to the Missouri Territory. Methodists there first met in 1815, in the same region where Nicholas Trammell settled along the White River. William Stephenson was initially appointed to the circuit, but the first church services were held in the home of an assistant local preacher by the name of Eli Lindsey. Lindsey moved from Kentucky to Lawrence County, Missouri, to live where Big Creek flowed into the Strawberry River. Rev. John M. Steele reportedly received the following firsthand account of an early church service near the White River settlements.

> Eli Lindsey began preaching on Strawberry in 1814 and his circuit ran from Little Red river north to what is now Missouri. He was a Methodist, and was said to have "his own methods." Colonel Magness states that the visits of Eli to Oil Trough Bottom were irregular; that he attended all the house raisings, log rollings, quiltings, marriages and frolics. He would encourage the young people to dance and after they were through would preach to them. At the end of the year 1815 he reported ninety-five members in his circuit. In 1816 he visited the spot where Batesville now stands and found a man named Reed in possession of a new house he had just finished for a store. Lindsey asked the privilege to christen it, which was granted. He sent out to Miller's creek, to Lafferty's creek, to Greenbrier, and all around, and notified the folks to come out. They came with their guns and a fine old crowd it was. Colonel Miller and his boys. Colonel Peel and Sons, the widow Lafferty and sons, Major Robert Magness and his army of boys, the Craigs, Ruddells, Trammels, Beans, Gillets, Holabys, Trimbles and Kelleys were all there with their guns stacked around the walls. Old Eli began his sermon and in a short time the dogs started a bear. Old Eli said: "The service is adjourned in order that the men may kill that bear." They rolled out with alacrity, mounted their horses, pursued Bruin and killed him. They then went back to the new house, where Eli "thanked God for men who knew how to shoot and for women who knew how to pray," and finished his sermon.[37]

Although the ending of that story seems crafted for the retelling, the list of attendees was likely not. Which Trammell kin were in attendance is uncertain, but the names of others associated with Nicholas Trammell are notable, and the likelihood that Nicholas Trammell was present at the service seems well founded.

Nicholas Trammell attended early Methodist gatherings in the Missouri Territory, but it is difficult to imagine his being overtaken by the religious fervor that prevailed during the Great Revival of 1800 back in Logan County, Kentucky. His tendency to hide quietly in the fringes where lawlessness and citizenship overlapped was one of the steadfast traits of his personality. Perhaps the religious zeal in Kentucky led to a backlash against his family pursuits of gambling and racing horses and forced him farther into the frontier. The reasons he moved will never fully be known.

As this story suggests, the Trammell family was not so far removed from daily religious life that they completely avoided the preacher. In fact, there is evidence that Trammell's connection to Methodism continued throughout his life. Trammell's second marriage was administered by an elder of the Methodist Church in Arkansas. Mote Askey's wife, Lucinda Hill Askey, was a devout Methodist. After Mote and his wife died, Nicholas Trammell raised their son, Harrison Askey, from a very young age. Harrison was later described at the end of his life as a Methodist, a Mason, and a Democrat in Gonzales County, Texas. Trammell was probably not irreligious, but a life in strict adherence to scripture was not compatible with the life of a trader and smuggler. Even a principle as basic as the Golden Rule would have been bad for Nicholas Trammell's business.

One Foot in Arkansas and the Other in the Red River

Nicholas Trammell's trading ventures coursed from the White River to the Red River, through Nacogdoches to Natchitoches, and possibly to Chihuahua and Santa Fe. In 1816 he was still using his land on the White River in the Missouri Territory as home base.[38] The Askeys were there as well, but the Mauldings stayed behind in Kentucky and Illinois. James Byrnside and Nathaniel Robbins, both from the Tennessee-Kentucky border, were White River residents as well. Both later operated ferries at key road crossings in Arkansas and Texas—Byrnside on the Red River south of Fulton, and Robbins at the Trinity River west of Nacogdoches, the same post that Nicholas Trammell would later hold. The Laffertys and the Musicks were traders from the same region who moved into southwestern Arkansas with the advancing immigration.[39]

Trammell's ties to Kentucky and Tennessee continued in spite of his relocation to the Missouri Territory. His dispute with Judge John Overton over his father's land certificate required Trammell's presence back in Kentucky in April 1814 to take depositions from Wesley Maulding, Morton Maulding, and Ragland Langston.[40] Almost twenty years passed between the date of the original grant and the final adjudication of the lawsuit with Overton in 1815. After an initial victory for Trammell, Overton appealed to the Supreme Court of Tennessee and won his appeal. By order of the court, half of the 640 acres in dispute were given to John Overton, already one of the largest landowners in the state and, by the time of the decision, a judge on the Superior Court for Tennessee. Both the circumstances of the matter and Overton's standing led to a loss in the courts for Nicholas Trammell.

That case was one of the only losses Trammell experienced when he pursued matters to higher courts. The loss was probably less consequential, since Trammell gained legal possession of his new 640-acre claim in the Missouri Territory in 1817 after a ten-year wait. Charles Kelly, one of the oldest settlers in the area and the man who became the first sheriff of Lawrence County, Missouri, proved Trammell's claim in Missouri when he testified that Trammell lived on the land for at least ten years.[41] True to his nature, Trammell's personality was not yet conducive to settling. His claim in the Missouri Territory may have been simply a place from which to advance his trading and smuggling activities farther south.[42]

Another trader of some notoriety visited the Missouri Territory in 1816—Jean Lafitte. The famous pirate was part of an expedition up the Arkansas River organized by an engineer named A. L. Latour. Latour pretended to be on a geographical survey, but the real purpose was a secret mission on behalf of the Spanish government to learn how some of the territory lost in the Louisiana Purchase might be regained. Part of their assessment was a military one. Lafitte and Latour made note of the many routes by which US military forces could cut off Spanish territory. Their findings did not comfort the authorities that engaged the mission. One of the strategic routes into Spanish Texas noted by Lafitte was the clandestine trading trail called Trammel's Trace.[43]

Lafitte's party ventured up the Arkansas River and stopped at the Arkansas Post where traders sent bear oil, tallow, salted bison, and pelts downriver to New Orleans.[44] Lafitte and Latour remained at the trading post for several months to gauge the temperament and loyalties of the traders, particularly those who ventured west to Santa Fe. Lafitte called himself "Captain Hillare" during their endeavor but fooled few people. His reputation in New Orleans traveled upstream with the trade. Lafitte and Latour heard complaints from the settlers toward the United States that could be construed as supportive of a return to Spanish rule. In spite of sentiments, the extension of the US government into its new territory in the form of trade and postal services made the transition to US rule a *fait accompli*.

The volume of trade Lafitte and Latour witnessed must have surprised them. Traders moved guns up the Red and Arkansas Rivers, exchanging them with the Indians for pelts and horses. Carbines purchased in Kentucky for $15 were exchanged for 250 pounds of fur, which could then be sold for $100 in New Orleans.[45] The profits were tempting, even if the resulting trade armed the aggressive Osages and incited further conflict among the tribes.

It would not be surprising to consider that Trammell and the famous pirate Lafitte crossed paths on Trammell's home turf. The two were connected many years later in an old legend of silver lost in Hendricks Lake at the Sabine River crossing of Trammel's Trace.[46] There were many Lafitte treasure legends across the south. If Laffite investigated the level of trading activity in Arkansas in 1816, he was no doubt aware of Nicholas Trammell's methods and knowledgeable of the clandestine roads Trammell used to move goods between the Red River and Nacogdoches.

Latour and Laffite hired crews to pole their keelboats upstream beyond the Arkansas Post, past Little Rock and Cadron (to the northwest), to the point where Fort Smith, Arkansas, was established in 1817. Although a strong US military presence was intended to stem Indian hostilities, the formula also included an attempt to make the Indians dependent on government trade goods to ensure their loyalty.[47] Trading posts, known as factories, were established near Fort Smith and in southwestern Arkansas on the Sulphur Fork of the Red River at Factory Bluff. The US factories sought to eliminate foreigners and unlicensed traders from doing business with Indians. For a few years between 1805 and 1822, this government-sponsored system competed with unsanctioned traders like Nicholas Trammell for the Indians' business in furs and hides.

The Sulphur Fork factory was a short-lived post, in operation only between 1818 and 1820. John Fowler built a post high atop Factory Bluff where the Sulphur River merged with the Red River. Fowler hoped to trade with the Caddo in that region and across into Spanish Texas on both sides of the Sabine. After only a year, Fowler saw that he was losing Caddo trade to the Spanish in Texas and losing pelts to unlicensed Natchitoches and Pecan Point traders like Nicholas Trammell, who provided the Indians with whiskey: "I think it proper to inform you that there are several white persons settled in the Cherokee Village (opposite the old Delaware Village and on the west bank of the Red river) for the purpose of trading with Indians. These men have no Lisence to trade and do it in direct violation of the Laws—It is thought many thefts are committed on the neighbouring whites under cover of the Indians."[48]

The illegal traders siphoning off Fowler's sanctioned business set up opposite an old Delaware Village on the west bank of the Red River. Just south of this location was the crossing site of Byrnsides' Ferry, one of the first across the Red River. Fowler's indictment described the same method of deception of which Trammell was accused by his Cherokee neighbors on the White River. Using the limited authority of his post, Fowler requested military help for the removal of the Red River whiskey traders, but as soon as troops took action with one group of traders, they were quickly replaced by others ready for commerce.[49]

The numbers of pelts Fowler took in trade put into perspective the incredible amount of game killed for commercial purposes. In 1818 and 1819 alone, Fowler sent 30,000 deerskins, 691 bearskins, 455 beaver pelts, and 1,183 otter, fox, raccoon, and wildcat pelts down the Red River through Natchitoches to New Orleans.[50] Like the eastern states, the new territories were quickly becoming overhunted and the game decimated. These huge numbers, for a factory that was for all practical purposes a failure, were an indication of why hunters from all nations lamented the loss of game and fought over rights to hunting grounds.

Roads Growing in Importance in 1819 Arkansas

The increasing numbers of people coming to Arkansas for trade created additional pressures for more improvements in the form of roads across the territory. Arkansas became a territory separate from Missouri on March 2, 1819. The very next day, the US Congress authorized an extension to the network of trails that crossed the new territory from the northeast near Batesville, Arkansas, to the southwest corner at Fulton and the Red River. The importance of roads to the development of the new territory escaped no one's attention. The involvement of Trammell and his extended family members in the emergence of early trading trails demonstrated their understanding that roads were the key to all forms of commerce.

The meaning of the word *road* must be understood in the context of the times. Conditions made travel difficult and confusing, even twenty years after the Arkansas Territory was formed. Trees cut close to the ground allowed wagons to pass over but left perilous stumps twelve to eighteen inches high. Blazes marked on trees along a route pointed the way for those who traveled the road for perhaps the only time in their lives. Since there was to be only one main road, there were often disputes about the best route. Everyone wanted the road to pass near their home. Settlers cut side roads resembling the main path and put their own blazes or marks on trails that led to only a single-family cabin. Misled travelers often found themselves miles off course. The farther south one traveled on the Southwest Trail, the worse conditions became.[51]

The Great Southwest Trail, known at the time as the National or Military Road, received attention and resources as the United States pushed its ever-widening influence into the western territories. Although the trail was used for years by early frontiersmen, later improvements opened it up for wagons and heavier migration. Soldiers with land grants earned for their military service pushed southward into Missouri and Arkansas after the War of 1812 ended in 1815. The military bounty warrants did not grant a specific place but instead provided a process for veterans to receive a patent to select public land. These warrants could be transferred, and many veterans sold their patents in a brisk trade with land speculators. Legislators opened up the frontier land to former soldiers because they believed battle-hardened war veterans might provide a buffer between the United States and the Indians being displaced even farther west.

Road building and trailblazing was not just an interest of the Trammell family; it was an obligation of the local governments that was diligently pursued. Nicholas Trammell's maternal relatives, the Mauldings, continued their engagement in the development of roads back in Kentucky. Trammell's uncle Richard Maulding was appointed in February 1817 to help maintain the Carmi-Kaskaskia road that ran across Hog Prairie to connect with the road from Carmi to Ten Mile Prairie, just inside the eastern border of Franklin County, Tennessee. In June 1819, Ambrose Maulding's son, Ennis, was appointed overseer of a road from "two miles west of Proctor's to Ambrose Maulding's."[52] In places like Kentucky and Tennessee, there were more improvements to the roads than in the Arkansas Territory as a consequence of their earlier settlement. Arkansas travelers were lucky to have a rough path. Back in Kentucky, Ennis Maulding's charge was to open the road to thirty feet wide, build bridges over all creeks and branches, and build log causeways across all muddy lands. These kinds of improvements would not be seen in Arkansas for many years. Road-building specifications at about the same time from the Circuit Court of Hempstead County, Arkansas, simply ordered roads laid out "the most direct and best way" from one settlement to another. Even that broad instruction was subject to alteration by the realities of the landscape.

A Boundary Defined—Adams-Onís Treaty

The year 1819 was a pivotal one in the sequence of historic events at both ends of Trammel's Trace. Settlers who danced around the uncertainty over the border along the Red River at Pecan Point found themselves evicted from their land because there were competing interests. Downstream at the Great Bend of the Red River, lots were sold at Fulton in hopes of developing that location into a strategic commercial port on the Red River above Natchitoches. At the southern end of Trammel's Trace, filibusters again emptied Nacogdoches of its population and almost completely destroyed the town in another attempt at revolution against Spanish Rule.

Since 1816, the United States and Spain had sought to settle their territorial claims and negotiate a boundary between the two nations. It was not until the Adams-Onís Treaty was finally signed in Washington on February 22, 1819, that the boundary between New Spain and the United States was spelled out. The treaty granted Florida to the United States, and Spain retained title to Texas. The Sabine and Red Rivers marked the borders between Spain and the United States, and a large part of the land formerly in the Neutral Ground became part of Louisiana and the United States.

In 1819, resistance to the Adams-Onís Treaty's cession of Texas to Spain coalesced into another organized rebellion to conquer that territory, led by James Long, a doctor and merchant from Natchez, Mississippi. Long advertised for people to join his militia in return for the promise of land. About three hundred paid "subscribers" to the rebellion occupied Nacogdoches on June 23, 1819 and declared Texas an independent republic. One of Long's cohorts, Hamlin Cook, traveled to Pecan Point in August 1819 to secure support at the northern boundary and to recruit more soldiers for Long's rebellion in Nacogdoches. Cook represented himself as being commissioned by the newly established Supreme Council of the Republic of Texas, a falsification of authority that did little to impress the residents there. Long's promise to establish order and offer land in his new nation was hollow and resulted in few recruits. Most of the Pecan Point settlers remained behind to tend to more important affairs—the harvest of a good year of crops. The independent Red River crowd was not easily swayed by promises from distant governments.

Long was emboldened by his occupation of Nacogdoches and sought to further his cause by securing the port of entry from the Gulf of Mexico at Galveston. Long traveled to Galveston to secure the help of Jean Laffite, but Long's absence from Nacogdoches was a setback for the rebellion. Not only was Long unsuccessful in getting Laffite's support, but while he was in Galveston, Spanish troops regrouped and chased his followers back across the Sabine River into Louisiana on October 28, 1819. Not long after Spanish troops reoccupied Nacogdoches, they captured two Americans and an Indian from Pecan Point who wandered unwittingly down Trammel's Trace into town on November 2, 1819. By that time, the settlement of Nacogdoches was ruined and emptied once again.[53]

The establishment of a formal boundary between Spanish Texas and the United States created a quandary for the settlers at Jonesborough and Pecan Point. Where for years they could claim residence in either the United States or Spain and parlay those nations' differences to their advantage, soon there would be no question where they lived. Some who had been at the Red River for a few years were eager to move deeper into Texas, nearer the commercial activity along the El Camino Real to San Antonio. The earlier uncertainty of the boundary was a magnet for trouble and an attraction for those who wanted to avoid the confines of the law. As one of their own noted, "We are a motley crew, emigrants from all parts of the world, and of course have all kinds of people, good and bad. But the bad seems to predominate."[54]

Spain now actively sought to secure its northern border, and the settlers realized their border-splitting society was coming to an end. "We hear that since the late movements of the Royalists in Texas, the families at Pecon Point on Red River, have re-crossed that river to Hampstead county, fearing an attack from the Spaniards."[55] The settlers' movement back to the north side of the river did not solve their problem of gaining legal authority to any land. Many Americans responded angrily to the Adams-Onís Treaty. The seemingly preordained right that many believed the United States had over the continent would not be easily interrupted by diplomatic consent between nations. More practically, settlers encroaching on both the Red

River and the Sabine could no longer come and go with the same freedom they enjoyed before the treaty.

Trammell's secretive trading ventures down Trammel's Trace were harshly interrupted. Events in the Spanish borderlands were a response to tensions over the borders between Spain and the United States and the increasing pressures from the young republic to the north to expand its territory even farther. Branches of Trammell's Trace from Pecan Point/Jonesborough and the Great Bend of the Red River were the lone roads into Texas from the north, and from that direction Texas soon felt increasing pressures of migration and change. Nacogdoches was in ruins, and the resultant military response to the Long filibustering expedition increased the likelihood that smugglers' goods and horses would be confiscated. The rules in Spanish Texas were changing, and it was time for Trammell to make a move.

6

1820–1826
Gone to Texas and Back

> *Every new country, when first opened to emigration, is settled by a strange mixture of heterogeneous elements— by the enterprising and the virtuous seeking to improve their condition, and by the vicious of different grades who desire to escape from the trammels of the terrors of the law.*
>
> —CHARLES SUMMERFIELD[1]

Although Trammell's ties to the Tennesseans around Pecan Point were well established, indications are that his ventures there were likely for trade and smuggling, not for farming or settlement. His home base was still in the area of the White River in northeastern Arkansas. His presence there was evident in the records for the sale of his land and his service on the first county court for the newly formed Independence County, Arkansas.

In order to lay claim to land in the Arkansas Territory, immigrants were required to improve the land and occupy it continuously for ten years, an act Trammell completed in 1817. As a testament to the pace of change in the region, the 640 acres Trammell settled on the White River changed jurisdictions from Lawrence County, Missouri, to Lawrence County, Arkansas, and finally to Independence County, Arkansas, in 1820. Trammell sold his White River land in 1821 to Morgan Magness.[2] Magness was the richest man in Independence County and had been in the region almost as long as Trammell. Trammell's sale of his land was a precursor to making his move toward Texas. Just as he had done with his move from Tennessee to the White River, Trammell had one foot firmly planted in each locale before making the move more permanent.

When the first petit jury for Independence County was formed by the Court of Common Pleas on November 19, 1821, Trammell was impaneled as a juror to try the case of Edward Sullins v. Arnold Schlasinger.[3] Since Trammell served on the jury in November of that year, it can be assumed that the sale of his land, which would have disqualified him from service, took place after that date. Trammell's involvement in the business of the court continued into the following year but on the other side of the jury rail. In January 1822, on the second day of court's new term, the judge discharged bail posted by Nicholas Trammell and an unnamed Askey, probably Mote, in the matter of Benjamin Freeman v. Nicholas Trammell. Trammell and Askey posted bail in this suit for an unspecified civil disagreement. Whatever money Trammell had tied up in the court was returned and became part of his assets to make the move toward Texas.

1820—Dewees and Dillard

Nacogdoches had its share of conflict, but it was not the sole epicenter of troubles in Spanish Texas. Although Nacogdoches was at times beyond control, it was at the Red River settlements of Jonesborough and Pecan Point where events best illustrated the difficulties of the time. Ever since the earliest settlers to Pecan Point came to capture mustangs and trade in buffalo hides, residents moved from one side of the Red River to the other, from one country to another, in response to Indian predations or the pendulum of political change.

Those issues, and the simple fact of the distance, made Pecan Point ungovernable. That did not mean that governance was not attempted. In 1819, Hempstead County, Arkansas, was expanded to encompass Pecan Point up the Red River. The organized government provided a mechanism for legal redress, but the courthouse was about one hundred miles and several days travel from the settlement. The Arkansas Territory created "old" Miller County in 1820, taking in most of what are now Bowie, Cass, and Red River Counties in Texas, shortening the distance a bit but complicating the legalities.[4]

Although Arkansas created this new political entity after the Adams-Onis Treaty in what was arguably Spanish territory, it also was part of land ceded to the Choctaw by the United States—a predictably troubling overlap. The Treaty of Doak's Stand in October 1820 set aside land for Choctaw Indians west of a line from Fulton to present-day Morrilton, Arkansas. Three hundred Anglo families north of the Red River who invested great effort clearing land and raising crops were understandably upset when the US military came to evict them. When the US government began to dislodge the settlers, the commanders may have assumed they would heed the warning and return to their previous homes in the eastern states. Some did return to the East, but many hoped their long-delayed transition from illegal squatter to landowner would not be denied. Migration south from Pecan Point into Spanish Texas began with a trickle in 1820. By 1822, a movement of people with historic impact commenced down a road from Pecan Point to Nacogdoches, widened for that purpose by Nicholas Trammell.

Two young men in that early migration left Pecan Point on foot in the late summer of 1820 in search of opportunity in Texas. William B. Dewees came to the Red River from Kentucky, and his traveling partner, Nicholas Dillard, was one of the many Tennesseans who settled there.[5] Provisioned with nothing more than blankets, rifles, and a little bacon, they traveled the "old Trammel trace leading from Pecan Point to Nacogdoches" during a time of year poorly suited for traveling through the woods.[6] When they were on the trail only three days, it began to rain incessantly. The rain continued for over a week and flooded everything around them. Dewees and Dillard quickly lost their bearings when creeks and rivers ran out of their banks and signs of Trammel's Trace were flooded. Their gunpowder was wet and unusable, and they ran out of food. In desperate straits, they struck off in the general direction of Nacogdoches and hoped for the best: "We were in a country where we were entirely unacquainted; with no road, no compass, and on the point of starvation. Our case seemed to be almost hopeless, but we still kept on, determined if possible to find relief, and travel as long as we were able."[7]

Dewees and Dillard came upon a hunting party of Caddo Indians who fed them and took them to a nearby camp, where they remained for four or five days. The Caddo resupplied them with food and gunpowder and provided directions down Trammel's Trace to Nacogdoches. The Caddo camp was four or five days of travel from the Sabine River crossing of Trammel's Trace.[8] On the north bank of the Sabine River, they met some French traders, a few Anadarko Indians, and two families of Americans who were camped and waiting for the floodwaters to recede so they could move southward. Dewees and Dillard acquired two horses from the Indians there and remained on the edge of the Sabine River floodplain for two to three weeks before the water receded enough for them to ford the river and continue their journey south to Nacogdoches. Prior to their arrival in Nacogdoches, however, a series of events took place that changed Texas history forever. They quickly returned to Pecan Point with dramatic news.

1820—Moses Austin to Texas

The movement of people into new territory was driven by something other than simple reason and at times seemed to defy it. Moses Austin, a businessman who was the father of the man later called the "Father of Texas," Stephen F. Austin, sensed that drive and acted on an idea in response to westward migration. Moses Austin built his wealth mining lead in Virginia, but his most recent business activities had failed. His visions for recovery led him farther south into Spanish Texas. In 1796, Moses Austin observed a pattern of illogical behavior in the settlers who originally moved to Kentucky that he kept in mind for future opportunities: "Ask these Pilgrims what they expect when they git to Kentuckey the Answer is Land. Have you any? No, but I expect I can git it. Have you anything to pay for land, No. Did you Ever see the Country. No but Every Body says its good land. Can any thing be more Absurd than the Conduct of man, here is hundreds Traveling hundreds of Miles, they Know not for what Nor Whither, except its to Kentucky, passing land almost as good and easy obtained, the Proprietors of which would gladly give on any terms, but it will not do its not Kentuckey its not the Promised land its not the goodly inheratance the Land of Milk and Honey."[9]

The name of any other western region settled in the 1800s could be substituted for Kentucky and the same tendencies applied. Movement westward was toward land unseen into circumstances poorly understood and largely without resources or wherewithal. Moses Austin's business sense led him to take advantage of these trends when he convinced the Spanish authorities to allow him to colonize Texas with carefully selected immigrants of reputable character.

While Dewees and Dillard were flooded on Trammel's Trace, Moses Austin traveled through Natchitoches and Nacogdoches on his way to San Antonio in search of opportunity on a grand scale. Austin arrived in San Antonio on December 23, 1820, where he sought permission to bring three hundred Anglo colonists into Texas. Austin had been planning this venture for some time. Moses' son, Stephen F. Austin, moved to Long Prairie in southwestern Arkansas in April 1819 to help facilitate the colonization project. Austin believed that a post near the boundary line would be useful to furnish provisions and provide a stopover for immigrants on their way to Texas. Initially, Moses Austin's request was refused by the Spanish governor, Antonio Martinez. A chance encounter with Baron de Bastrop, an old acquaintance of Austin's who knew the governor well, resulted in a second meeting and the approval to establish an Anglo colony in the interior of Spanish Texas along the Colorado River.

Moses Austin's trip through Texas by way of Nacogdoches had already created a stir, even before the authorization was granted. Speculation that a colony was being considered quickly transformed into detailed rumors of its imminent fact. José Erasmo Seguín, who assisted Austin with his colonization effort, offered his observations to the governor regarding the increasing migration that took place even before Austin was formally granted authority to bring in colonists. "It seems, when Austin passed through this place [Nacogdoches], he considered as granted the authorization to come with 500 families to the Colorado River. The news went abroad; and, the people of Missouri, who are admitted, as well as those who were not, such as the people of Pecan Point, have taken the advance and built their houses, from the Sabine down to Nacogdoches, and even farther as I am informed. . . the families are large and poor and have no means of transportation."[10]

An interesting aspect of Seguín's letter to the governor is that it reveals that there were specific judgments made about who was desirable and permitted to settle in the new colony and who was not.

The settlers from Pecan Point would not be welcomed in Austin's colony.

Their character was widely discussed in letters among government officials. That reputation was repeatedly reflected in official communication and government decisions about migration. They were described as bad men, desperados, and renegades, and that reputation carried more weight than any facts to the contrary.

When Dewees and Dillard arrived in Nacogdoches, news of the purpose of Austin's trip to San Antonio gave the tentative outcome a sense of certainty. They quickly reversed course up Trammel's Trace and returned to Pecan Point to prepare for their permanent relocation to Texas. Although Trammel's Trace would soon be busy with Red River settlers moving south into Texas, all traffic briefly came to a standstill after the news of Austin's colony spread. On their return trip from Nacogdoches to Pecan Point along about two hundred miles of Trammel's Trace, a camp of Delaware Indians were the only humans Dewees and Dillard encountered on the entire journey.[11]

1821, July—Austin Enters

The difficulties of a four-week journey back out of Texas during the winter months left the fifty-six-year-old Moses Austin malnourished and in poor health. Although Moses learned his "Texas Venture" was moving forward, he never recovered from his exposure to the weather. Moses Austin died at the home of his son-in-law, James Bryan, on June 10, 1821. His dying wish two days before his death was for his son, Stephen, to continue in his place. Stephen F. Austin did not learn of his father's death until three weeks later, while he was on his way from Natchitoches to Texas along the El Camino Real. His father's death made Austin even more determined to fulfill their vision.

When Stephen F. Austin crossed the Sabine River into Texas on July 16, 1821, the history of the region began a new chapter. His first few days in Texas were a fitting introduction to this new country. At the Arroyo Hondo west of Natchitoches, Austin's path crossed between two large magnolia trees whose bark was etched with the names of others gone before.[12] On his first night in Texas, Austin slept along the El Camino Real at a longstanding campsite used by smugglers who moved horses and contraband back and forth between New Spain and the United States. Upon reaching the land around Borregas Creek, just west of modern-day Milam in Sabine County, Austin wrote in his diary that it was "covered with the most luxuriant growth of Grass I ever beheld in any county." For breakfast on Tuesday, Austin stopped at the home of William English, whom he later included in a cadre of "bad men" when English supported another uprising in Nacogdoches. On his third day in Texas, only twelve miles east of Nacogdoches, Austin and his party overtook illegal traders moving mules and horses across the border, stock stolen by the Comanche. Austin was confronted with situations just like these for years and, just as unsuccessfully as others before him, attempted to prevent such trade and trafficking. By his fourth day in the new country, Austin entered Nacogdoches. At the time it was but a rundown village, yet it was still the only Spanish presence east of San Antonio. Seven houses and a stone structure used for Indian trade were all that remained.[13]

It did not take long after his arrival in Texas for Austin to notice the volume of illegal trade. When he submitted to the captain general of the North, Anastacio Bustamante, his recommendations for what should be done, Austin correctly identified the trouble spots through which horse smuggling took place—Nacogdoches and Pecan Point. Through these two exit points to the United States, traders could bring in goods and receive horses and mules in trade.¹⁴ Both of these exit points for smuggling out of Texas were served by Trammel's Trace. Nacogdoches was a key hub, providing access to willing buyers in the United States. Pecan Point on the Red River was a primary source for mustangs and buffalo hides. The route to the north was up Trammel's Trace and across the Missouri Territory. Indian agent Sibley had recognized these conditions ten years earlier: "Besides the Encampment of Robbers on this Side of the River Sabine (near Nacogdoches), there is another Collection of Bad Men & Some Women at the Pecan Point on Red River, a Most Beautiful Place."¹⁵

Early immigrants were not always those with the upstanding character Austin wanted for his colony. Texas was a magnet for outlaws, land speculators, and settlers inspired by hopes of riches in the vast new territory. Even though the persistent Red River settlers had been evicted by the US military, they took little time in resettling the region. William Styles, the object of General Bradford's ejection from the Kiamichi River in 1819, was still there in 1821 despite the expulsion. Spanish efforts on their side of the Red River were ineffectual and did little to change the course of trade. The already porous borders of Spanish Texas were flooded from the outside, applying even more pressure on the volatile circumstances. The policies of nations toward settlements thousands of miles away had little impact on behavior, and no settlements were more distant than Pecan Point and Jonesborough.¹⁶

Threats from beyond their boundaries were not the only worries for the Spanish government. On August 24, 1821, a movement that had fought for Mexican independence since 1810 succeeded in wresting control from Spain. Spanish Texas became Mexican Texas, and Austin's colonization contract with the Spanish government was suddenly void. After extended negotiations and personal distress, Austin was able to renegotiate his contract and keep colonization moving forward. In fact, the prevailing view was that the new Mexican government would be more liberal in its policies on immigration. This perception only served to increase the movement of people from the United States into Texas, especially among the prohibited settlers at Pecan Point.

Their hopes of securing land claims seemed to improve, and many Red River families headed south into Texas along Trammel's Trace. Former Americans settled at the Red River wrote a letter to the Mexican governor of Texas. Their 1821 letter complained about the ongoing illegalities on the north side of the river: "The settlers on the north side of Red River carry on direct trade with the Comanches furnishing them with all the munitions of war and receiving in exchange a great number of horses many of which bear the Spanish brand. We feel this selfish and illegal traffic is very injurious to your government."[17]

Not surprisingly, communication about troubles with thievery, smuggling, and illegal trade followed Nicholas Trammell wherever he went. Although he lived in far northeastern Arkansas, his trading ventures and kinships with the Red River settlers offered him ample opportunity to be a part of the illegal trafficking going on at Pecan Point.

Confusion over the jurisdiction of two nations allowed opinion and circumstance to establish citizenship almost at will. On the south side, although they clearly understood that they lived in the Spanish Province of Texas, settlers also wanted the protection of the laws of the United States. And even though the residents on the north side of the river had been under the jurisdiction of the United States, their land was no longer legally accessible to them as a result of the Choctaw treaty. A confluence of competing interests and jurisdictions ensued. Whether any settler along the Red River lived in Spain, the United States, or in Indian Territory remained unsettled. Residents on both sides of the Red River were troubled and unsure. Austin's colony seemed like the answer to their uncertainty.

1821—"Bad Men" Enter Texas

If Austin made a conscious decision to exclude Pecan Point settlers from the proposed colony, as José Erasmo Seguín's letter suggested, it must not have been clearly demonstrated to the settlers themselves. Shortly after Mexican independence and before Austin could secure his colony, many families quickly moved farther into Texas in hopes of their presence being legitimized. Several of the Red River families became part of Austin's "Old Three Hundred" original settlers. By June 1821, several Pecan Point families were preparing to make the trip to the new colony, even before receiving any assurances of what it was they were heading toward. On the first day of the new year, January 1, 1822, William Dewees arrived at the edge of the new colony along the El Camino Real west of Nacogdoches. "After a long and toilsome journey I arrived at this point from Red river, in company with three or four families from that country, on the first day of January, last. We encamped at the crossing of the old San Antonio road two miles above the mouth of the Little Brazos river. We were several months in getting here, there being several families in company, besides this, we lost several horses on the way, and in fact we seemed to meet with a great many misfortunes. We carried our luggage entirely upon pack-horses, the roads being perfectly impassable for a vehicle of any description."[18]

Two other early Red River immigrants who used the Southwest Trail and Trammel's Trace for their migration south to the new colony were Thomas Boatwright and Daniel Shipman. Boatwright and his kin migrated to Arkansas in 1819 and then to the Red River near Spanish Bluff west of Fulton. In the fall of 1821, Boatwright, his wife, and their ten children traveled Trammel's Trace to Nacogdoches. They followed the section of the Trace that connected the Red River settlements of Pecan Point and Jonesborough to Nacogdoches. In early December 1821, Boatwright left Nacogdoches for Central Texas, where he received grants of land in Austin's new colony.

Daniel Shipman and his family arrived in Jonesborough the following year on March 19, 1822, with plans to continue their journey farther into Texas to Austin's colony. Shipman was an unsettled type. Between 1814 and 1822, Shipman and his family farmed in North Carolina, Tennessee, Illinois, and Missouri and then finally came to the edge of Arkansas' boundary with Mexico, poised for their next opportunity. After only four days' rest in Jonesborough, Shipman and a companion began the trip south. "When we left Jonesborough, we were told to go down the river to what was then called 'Pecan Point,' a neighborhood about twenty-five miles below Jonesborough, and there inquire how to get to the Nick Tramel trail, which we did and got along very well."[19]

At the Sulphur Fork (later known as Stephenson's Ferry), Shipman made a difficult ford across a deep stream and encountered a small village of Alabama Indians. Since earlier travelers along the same route noted no human settlements of any kind, this was likely a temporary camp for tribes moving through the area. If there were Indians in the area for any reason, they certainly noted the increasing traffic down Trammel's Trace.

1820–1822—Indian Troubles, Cherokee Settlements

Not only did the prospects of land lure the Pecan Point settlers south farther into Texas, Indian troubles also eroded the comfort they found along the river. Far from any other Anglo settlement, the residents themselves solely bore the threat of Indian attack. The Anglo squatters contributed to the unrest by regularly hunting in the land west of the Kiamichi River, where an act of Congress prohibited their entry into Osage hunting grounds. Not all interactions with Indians were hostile. Friendly trade and the relative civility of the Choctaw, Delaware, and Shawnee made relations less threatening most of the time. The Choctaw were among various bands of the "Five Civilized Tribes" that migrated into Texas in the last years of Spanish rule and the first years of Mexican jurisdiction. The Choctaw also acted as middlemen in the traffic of stolen horses along the Red River carried on along Trammel's Trace. Battles between the Choctaw and the Osage over hunting grounds had a disquieting influence on the region. In 1822, the Choctaw could not tolerate the hostilities of the Osage any longer. A splinter group of sixty Choctaw warriors who lived on the south side of the Red River began a string of raids against the Osage on the north side. The Osage retaliated, not only on the Choctaw but on the Anglo residents as well. The Osage also continued to steal horses from mustang hunters. Both settlers and Choctaw contributed to the "continual state of apprehension."[20]

Trammell and the Cherokee

The Cherokee of that region and Nicholas Trammell shared a common migration history across three decades. The traditional Cherokee homeland was east of the Kentucky-Tennessee border, where the Trammells once made their home. Cherokees sent annual hunting parties as early as 1770 to the Arkansas and White Rivers on the western side of the Mississippi, where Nicholas Trammell later lived. Cherokee traders traveled down the Red River by 1807 to trade in Natchitoches. Cherokee accused Trammell and Askey of theft in 1813. A military party in April 1817 found Cherokee villages on the east bank of the Red River near Factory Bluff at the mouth of the Sulphur River. By 1819, when the Arkansas Territory was separated from Missouri, there were over three thousand western Cherokees seeking land in the same regions as Anglo settlers and hunting in grounds claimed by other tribes.[21] Conflicts from all sides left the Cherokees ill at ease. Overlapping interests in trade and settlement no doubt led to many exchanges between Nicholas Trammell and his Cherokee neighbors.

A splinter group from the eastern Cherokees was particularly contemporary to Trammell in time and place. In the early months of 1810, shortly after Trammell's move to the White River, a Cherokee chief named Duwali (also known as Chief Bowles) moved seventy-five of his people into the Missouri Territory. As early as 1820, Cherokees in Arkansas complained about encroachment of Americans who regularly came into their territory. Nicholas Trammell is not named in this accusation, but it certainly reflected the tactics he had used before: "There was good reason to believe that white men had engaged the Indians for this thievery, purchased the stolen horses, and then ran them south. We think there can be little doubt that the white outlaws are the principal instigators of the depredations committed by the Indians."[22]

Chief John Jolly, a respected chief who adopted Sam Houston as his son years before, feared the thievery would upset the relative calm. Jolly was unable to stop the trade Chief Bowles's band of renegade Cherokees carried on with the Anglo horse traders. In a letter published in the *Arkansas Gazette* in 1820, Chief Jolly made note of his attempts to suspend the stealing: "In council yesterday we appointed three mounted men authorized to suppress all thefts of every kind. We wish to adopt all measures to keep the white men from carrying on trade with these Cherokees, which might disturb the harmony that exists between the white people and the Cherokees."[23]

After other tribes forced Bowles out of the region, he tried to establish a village at the Three Forks of the Trinity, near present-day Dallas, Texas. Yet again, stronger tribes forced his people out, and they moved to Lost Prairie in southwestern Arkansas. Life there became an uncivilized existence for the culturally adaptive Cherokee. Unlicensed traders cheated the Indians and sold them whiskey, making the villages "poor and miserable." Overhunting depleted the supply of game and pelts on which everyone relied. Anglo citizens of the newly formed Hempstead County, Arkansas, complained of these "strolling bands" of Indians in Lost Prairie. Bowles and other Cherokees were accused of stealing horses at Pecan Point, resulting in a request of the governor to return the Indians to the Arkansas River. Having moved from Petit Jean Creek near the Arkansas River to the Three Forks of the Trinity and then to Lost Prairie in only a few years' span, Chief Bowles and his people were weary of the conflicts. In the winter months of 1820, Bowles's Cherokees moved again, this time into the uninhabited forests west of Trammel's Trace about fifty miles north of Nacogdoches. There they built what became the largest Cherokee settlement in Texas.[24]

As word of the Mexican openness spread, Anglo settlers built cabins and cleared land for crops. Only those too hesitant and tentative for such a frontier life waited for official confirmation of their right to be there. Chief Bowles and his Cherokee leaders realized their comfortable settlement would soon be an area of interest for even more Anglo settlers. The Cherokee wanted title to their land. They understood that a land title was the closest thing to security they could have in a time when governments were changing and many nations had an interest. The Spanish government gave the Cherokee permits to live in the province but not clear titles. With Mexican officials in charge, even their uncertain permits were in jeopardy. Richard Fields, diplomatic chief of the Texas Cherokees, pled his case to James Dill, the alcalde at Nacogdoches, on February 1, 1822. The alcalde was akin to the local magistrate. "Dear Sir: I wish to fall at your feet and humbly ask you what must be done with us poor Indians. We have some grants that were given us when we lived under the Spanish Government and we wish you to send us news by the next mail whether they will be reversed or not."[25]

Dill sensed the gravity of the question and did not respond, so in November 1822, Fields and a small delegation set out for San Antonio. When Fields made his case for land in that city, the Mexican governor also begged off and said he had no authority to grant titles to land. The Cherokee were able to reach an agreement with the governor, but not for land, only for the promise of land. Article three of their agreement clearly stated the governor's interests: "That a party of the warriors of said village must be constantly kept on the road leading from the province (Nacogdoches) to the United States, to prevent stolen animals from being carried thither, and to apprehend and punish those evil disposed foreigners, who form assemblages, and abound on the banks of the River Sabine within the Territory of Texas."[26]

The governor of Texas, José Félix Trespalacios, wanted the Cherokees to be his local militia to combat the trade in stolen horses. While Trespalacios saw the Indians as a means to stop the trade in horses, Stephen F. Austin recognized that the Indians were among those most responsible for the trade the Mexican governor wanted them to help prevent. Austin wrote: "The large number of horses and mules that fell into the hands of the savages they exchanged for guns, ammunition, and whatever else pleased their fancy. The traffic in horses and mules became so extensive that well beaten trails led from the interior of the border provinces to the frontier of the United States, and it proved so lucrative to those engaged in it that the Indians were encouraged to prosecute their robbing and plundering expeditions against Texas, Coahuila, and Nuevo Santander with ever increasing ferocity."[27]

The Mexican plan was to have the local Indians enforce smuggling laws against the Anglos, with whom they were partners in stolen horses. Such action would have led to certain disaster. Austin knew the Mexican officials did not have a realistic plan to stop the trade, but his own ideas had not been successful in the past either. Austin believed that "much good was expected to result from simply blockading the trade routes by stationing garrisons, or by making settlements at suitable points."[28] That approach had not worked for the Spanish, and it also proved unworkable on the Mexican border with the United States.

Chief Fields returned to eastern Texas in June 1822, disappointed yet again. There would be no land ownership for the Cherokee, no titles affirming years of promises. In fact, there was mounting pressure from Anglos who came into the area between Nacogdoches and the Sabine to move the Cherokees out of the region. New settlements in the Neutral Ground along Ayish Bayou, on the Angelina and Sabine Rivers and near the Attoyac Bayou east of Nacogdoches, bounded the Texas Cherokees. Chief Bowles and his Cherokee people along Trammel's Trace were in a no man's land of their own.

1823-1824—Immigration from All Directions

The Mexican government was increasingly nervous about the migration into Texas that was growing beyond its control. Baron de Bastrop, the man who helped Moses Austin secure his colonization contract, was directed to investigate the growing numbers of Anglo immigrants from the Red River border down to Nacogdoches and provide a report in late 1823: "As your communication of 24 September did not come to hand until 10 November, I was unable to go to the Sabine River and from there to Pecan Point on the Red River of Natchitoches in order to personally inspect all the families that live in that area—the only way to give an exact accounting of them. From all the reports I have been able to gather, there are two hundred families between the San Jacinto and Sabine rivers, most of them natives of the country and settled there for many years, and fifty at Pecan Point."[29]

Bastrop's report was little more than a repetition of rumor about the numbers of people settling in Mexican Texas. The scope of the problem continued to escape the Mexican government for several years.

Undesirables from Pecan Point and exiles from the United States who were denied entry to Austin's colony occupied the former Neutral Ground to the east and north of Nacogdoches. To discourage further migration toward the interior, newer immigrants who crossed the Sabine into Texas and asked about moving past Nacogdoches were told stories of violence and anarchy by the squatters in the Neutral Ground, deterring their entry.[30] The Mexican government acted on the assumption that the presence of an undesirable element at the border served as a buffer for the more acceptable settlers in the colony to the west. This buffer mentality was a common, if flawed, belief during the early periods of settlement. The Spanish hoped the Indians deterred entry by Anglos, so they allowed tribes to settle in the boundary lands. In this Mexican adaptation, the authorities believed the degradation of character of those in the borderlands acted as a defense to further migration and might somehow prevent encroachment into Texas. Instead, the policy created a region of lawlessness and chaos served by access from the north down Trammel's Trace and an escape route across the Sabine to the United States on the El Camino Real. The government's belief in a buffer zone may have delayed the advance of some immigrants but certainly did not prevent those bold enough to make it to the border from continuing their quest for land.

Anglo settlers around Nacogdoches who arrived years before this most recent wave of immigrants were unhappy with the Mexican complacency and wanted to take matters into their own hands. James Gaines, a notably aggressive early resident, wrote a letter to the Mexican governor in August 1823 complaining that these new exiles from the United States were upsetting the "old settlers" by taking their land. In Gaines's view, the latest cadre of good-for-nothing foreigners was intruding into the Sabine District without permission. As an elected representative of the local citizenry, Gaines believed he was authorized to capture, judge, whip, and expel "six roguish families and ten straggling vagrants" on behalf of Mexican interests. True to his rather ruthless reputation, Gaines argued that another twenty families deserved the same treatment if property rights were to be made secure in the region.[31]

On the northern border between Mexico and the United States at the Red River, increasing conflict was spurred by US Indian policy. Conditions there ultimately led to a recommendation by Gen. Winfield Scott that the United States establish a military fortification in that region.: "The evils existing on our Spanish border. . . have been carried to a great extent. No question exists but that in addition to quarrels in that neighborhood among the parties of different tribes of Indians, the laws of the United States relative to the introduction of slaves and to trading with Indians are set at perfect contempt and daily and extensively violated. In addition to this, a band of lawless marauders have established themselves on the Red River, and are in the habit of committing the most outrageous acts of robbery, violence and murder."[32]

Soldiers chose a site for the new fort on Gates Creek about ten miles above the mouth of the Kiamichi River, and a small army contingent built Cantonment Towson in May 1824, near present-day Fort Towson, Oklahoma. This location was chosen to help regulate trade between settlers and Indians and to maintain the peace by stopping the hostilities. The Red River settlements were in turmoil. By 1825, Indians and Anglos occupied the same territory, for the same reasons, and with similar hopes to acquire land and hunt the prairies. There were several Indian villages along the Red River: one camp near Spanish Bluff, and villages of Delaware and Kickapoo along the prairies south of the river. Lingering delays in the Choctaw Treaty as a result of legal wrangling ended with a modification to the treaty boundary in January 1825. The Choctaw Line moved to what is roughly now the Arkansas-Oklahoma border. After that clarification, any residents north of the Red River knew they unquestionably lived in Choctaw territory.

The presence of the US Army accomplished little and angered everyone. Even Anglos attacked the soldiers when they tried again to evict settlers from their homes.[33] Fort Towson was abandoned in early 1829, and the remaining Anglos angrily burned the log structure to the ground.

The only other US military presence with proximity to Texas was far to the east in Louisiana. Two years before Fort Towson was built, the United States established Fort Jesup in 1822, near Natchitoches, east of Nacogdoches and the Sabine River. The fort was on high ground not far from the El Camino Real to help protect settlers in Louisiana and to monitor an unlikely invasion from Mexico. Besides the main fort, soldiers built a small fortification on the Sabine River, closer to Mexican Texas and not far from what became Gaines Ferry. Even with a military presence in close proximity to the road, soldiers at the fort did little to stop the flow of trade. A "way around" could always be found.

One of the most important outcomes of the construction of these two forts was the road between them. The Fort Towson to Fort Jesup road was on the US side of the Red River, arcing for hundreds of miles between Pecan Point and Natchitoches.[34] It facilitated trade and migration along the boundary. Between those two destinations, Trammel's Trace crossed the Fort Towson–Fort Jesup road and set off for Texas from two separate points, at Jonesborough/Pecan Point and Fulton. The military road did not lead to Texas, but Trammel's Trace did, and it was now becoming a busy path for immigration.

1822–1824—All Roads Lead to Texas

Where people traveled, roads followed. Improvements to the Southwest Trail from St. Louis, Missouri, extended it by 1820 into northeastern Arkansas as far as Poke Bayou (Batesville) and Cadron, Arkansas. By 1821, it extended through the villages of Washington and Fulton, Arkansas, to the Red River. Four-fifths of new arrivals into the Arkansas Territory after 1817 came by way of the Southwest Trail. Observations of the increased traffic through Little Rock led to a forecast of an immense immigration southward during the spring of 1822. The *Arkansas Gazette* reported in November 1822 that hundreds of people were moving south toward Texas from the region around Nicholas Trammell's former home along the White River.[35]

To get to Texas from Fulton, travelers either took Trammel's Trace or stayed on the eastern side of the Red River, down the road through Lost Prairie to Natchitoches, and from there to the El Camino Real.

Roads were open to anyone, but ferry crossings required payment. This commercial interest led to several ferry crossings being developed or improved. For a fee, travelers could avoid the typical problems of water crossings and move themselves and their property across by boat. James Byrnside, a contemporary of Trammell's from Long Prairie, was authorized in August 1820 to cross the Red with a ferry at the "Delaware Village." Byrnside's ferry was twenty-eight miles due south of Washington along a steep bluff—the same location where unlicensed traders established a post to trade with the Indians.[36] The business of helping settlers cross the Red River must have been brisk. The county authorized Alexander S. Walker to keep another ferry at the same location in March 1822 on the opposite side of the river and then authorized a third ferry, also on the west side.[37] A little farther south, another ferry was available at Musick Prairie, operated by Robert Dunn.[38]

Heavy traffic into Mexican Texas offered opportunities to many ferry operators. Other ferry licenses to A. Tidwell, William Slingland, A. S. Walker, William Talbot, Isaac Pennington, and William Stevenson were issued in southwest Arkansas between March 1822 and October 1823, providing many ways to cross the Red River both north and south of Fulton. By 1824, a ferry was authorized at Pigeon Bluff and, in the following year, another just above Chicaninny Prairie.[39] Bryan T. Nowlin kept two ferries, one near his house on the river and one at Fisher's Prairie, and by 1825 Daniel Cornelius kept one near his house.[40] Roads provided access to each of these crossings, maintained by the ferry operators in an attempt to guide traffic toward them. Tavern permits issued in conjunction with ferry licenses to Pennington and Byrnside were an indication of their ideas about what ferry passengers might do with the time they spent waiting for the ferry.

One particular Red River ferry was central to the history of Trammel's Trace, and one of its operators connected with Nicholas Trammell. In July 1821, the Hempstead County court authorized a ferry at Fulton operated by Aden Bunch and David Roberts.[41] Only a few months after being granted the ferry license at Fulton, Bunch sold all of his possessions and migrated down Trammel's Trace to Nacogdoches. In a strange twist indicative of the sometimes complex relationships within a small population of settlers, Bunch was later accused of murder by none other than Nicholas Trammell (a story told later in this chapter). With traffic toward Trammel's Trace feeding business on Bunch's ferry, it is likely that Bunch and Trammell knew each other before the ferry was established.

The road leading from Washington to Fulton at the head of Trammel's Trace continued to receive attention and occasional rerouting. In May 1826, Absalom Eden, William Shaw, and Elijah Stuart were assigned as commissioners to view and mark the road and reported the route they used at the next term of the court: "from Washington to intersect the mine creek Mount Prairie road, thens crossing Shaws creek, thence with the old choctaw line to Edens, thence to John Wilsons in the Bodark prairie, thence to the Town of Fulton."[42]

Road commissioners typically had a vested interest in their assigned road, either to redirect commercial traffic or simply for the convenience of travel to their homes. Roads were often abandoned due to poor conditions, a lack of use, or because a better route was available. This route was not one of those. It was well used for many years before and after the wave of immigrants heading toward Texas.

When hundreds of people headed south from the White River, Nicholas Trammell and his entire extended family joined in the migration to Texas. Trammell and others opened the road from the Red River to the forests of eastern Texas. The road he once used to move horses and trade goods was now the route he used to take his cattle, property, and family toward the opportunity for land in Stephen F. Austin's colony.

From Arkansas to Texas

Exactly when Nicholas Trammell relocated his usual collection of extended family toward Texas is not precisely clear, but there are indicators documenting his movements that suggest he was in Independence County, Arkansas, as late as 1821. If Trammell's prior habits surrounding his migration across the emerging territories held true, he left the White River in stages beginning in 1822 and was on his way south toward the Red River settlements around Pecan Point.

Nicholas Trammell moved to the Red River boundary between the United States and Mexican Texas by the time migration to Austin's colony began to increase in 1822. When Dewees and Dillard made their second trip back to Texas late in 1821, after learning about Austin's colony, Trammel's Trace was not suitable for wagon traffic. When the Trammells and other Red River families began their move into Texas soon after being alerted to the prospects of colonization, he widened the route with "chopping axes and hatchets" to allow families to move livestock and belongings down the trail.[43] These implements would have been brought from Tennessee, no doubt with a grinding stone to keep them sharp. By following existing paths, their work would have been easier, but Trammell and his men would have hand-cleared trees and brush. They could leave stumps above the ground but low enough to pass under a wagon. Horses or mules would have been used to pull aside downed tree trunks or to remove large stones. This was no inconsiderable effort but one that was undertaken by determined men . . . determined to head toward Austin's colony.

With a rough but widened road from Pecan Point connected to the more frequently used leg of the trail between Fulton and Nacogdoches, the entry way for Red River settlers to migrate into Texas was cleared. Two of the three entry points into Mexican Texas were at the Red River on branches of Trammel's Trace.

Stephen F. Austin and the Mexican government may have opened the door to Texas, but it was Nicholas Trammell who cleared the trails to get there. Trammel's Trace would never again be a clandestine trail for smugglers. It was quickly becoming a primary route toward the riches of land possible in Texas.

The preferred point of entry into Texas for many from Tennessee and Kentucky was not at the Great Bend of the Red River where Fulton was taking shape but at the trading settlement of Pecan Point farther upriver. The road from southwestern Arkansas to Pecan Point, and on to Jonesborough farther upriver, was established as early as 1815 over trails used by long hunters and traders from the territories to the east. The road to these two Red River settlements left Fulton and coursed westward past Mound Prairie, Arkansas. It followed the north side of the Little River to a low-water crossing, where the trail cut across toward the trading post at Pecan Point.[44] Trammell had no doubt previously used this route for trading in captured mustangs and stolen horses. Now he used it to begin the relocation of his family and possessions into Mexican Texas.

Looking at a map of the general course settlers from Tennessee and Kentucky used to get to the interior of Texas inspires questions. A solely geographical view of their migration by way of Pecan Point clearly shows that although settlers could have taken the traditional route of Trammel's Trace down through eastern Texas, they instead traveled much farther west to Pecan Point, Jonesborough, and farther upriver to the mouth of the Kiamichi River. Why would early settlers go so far west from Arkansas only to head back to the south and east to Nacogdoches? Why not simply head down Trammel's Trace from Fulton to Nacogdoches? The likely answer is that there was a significant early settlement of kinsmen from Tennessee and Kentucky already at Pecan Point by virtue of the trading posts that grew up at the long-time buffalo crossing. Austin's colony in central Texas was the focal point for the migration from Pecan Point, and the way to get there was down Trammel's Trace and then by way of the El Camino Real through Nacogdoches.

By the end of 1819, Nacogdoches was emptied of inhabitants and influence. Attempts by Anglo filibusters to use the settlement as a focal point for rebellion against Spain left it devoid of those who feared for their safety or who were complicit in the effort to overthrow Spanish rule. The old Spanish post at the southern end of Trammel's Trace was a desolate place with only a few remaining buildings. A limited number of Spaniards, a few French traders, some Americans returning from safety across the Sabine, and a contingent of free Negroes filled the void after the Long Rebellion. Smuggling continued unabated by any military resistance, and the only impediment to commerce along the El Camino Real and Trammel's Trace was the scarcity of customers.

Trammell's whereabouts from early 1822 to late 1823 are indeterminate, but his presence at the Red River and his role in widening the old road from Pecan Point to Nacogdoches are certain. Between 1822 and 1823, when it seemed all of Arkansas was emptying into Texas, Trammell and his family did the same. In 1824, his reemergence in court records in Nacogdoches answers the question of his whereabouts.

Indications are that Trammell arrived in the Nacogdoches District sometime in 1823. The first record establishing his residence near Nacogdoches was on January 16, 1824. Trammell signed a petition in support of the alcalde, James Dill, along with other notable Nacogdoches residents Peter Ellis Bean, James Gaines, and William Goyens and other "old settlers" of the region.⁴⁵ James Dill, an early Anglo immigrant to the region, was the only remnant of local governance after the Long Rebellion emptied Nacogdoches in 1819. Dill left during the conflict but returned to Nacogdoches in 1820 and became the first Anglo-American elected alcalde for the Nacogdoches *ayuntamiento*, or governing council. In 1823, Juan Seguín contested Dill for the post and lost, but Seguín, John Durst, and others removed Dill from office and installed Seguín on October 10, 1823. That Trammell was present in Nacogdoches when this political maneuvering took place is likely. By the time Trammell signed the petition supporting Dill early in 1824, Trammell had been part of an Anglo faction in Nacogdoches long enough to develop his alliances.⁴⁶ Evidence that choosing sides had inevitable consequences soon became clear. On the same day that Nick signed the petition protesting Seguín's installation as alcalde, Seguín issued a summons for Nicholas Trammel to answer for an unspecified debt to John York. Although he was in Nacogdoches, Trammell's connections to both the White River and Kentucky also lingered. In November 1824, Trammell made payment on a court judgment to a justice of the peace in Independence County, Arkansas, for a legal matter back in Livingston County, Kentucky.⁴⁷

When Nicholas Trammell widened the road south from Pecan Point, it was to allow his family and others to move their belongings by wagon to the interior of Texas, but Nacogdoches was not his chosen destination. Like many others, he sought land and fortune in Stephen F. Austin's colony in Mexican Texas. Trammell's eventual residence near Nacogdoches was a second choice—a result of rejection by Stephen F. Austin himself.

Austin Rejects Trammell

Stephen F. Austin was a particular man. On many occasions, for many different correspondents, he referenced his standards of entry regarding the character of his colonists. His letters are filled with clarity on his position. In 1823, he wrote "No frontiersman who has no other occupation than that of hunter will be received—no drunkard, nor Gambler, nor profane swearer no idler, nor any man against whom there is even probable grounds of suspicion that he is a bad man, or even has been considered a bad or disorderly man will be received." Later that same year, he more specifically told a partner, "you must examine the Red River emigrants very closely and take care that no bad men get in—let us have no black sheep in our flock."⁴⁸

Evidence that Stephen F. Austin met Nicholas Trammell in 1822 or 1823 and refused his acceptance into the colony is present in a letter Austin wrote in March 1826. Austin angrily wrote to Governor Saucedo after a man named Edmond McLocklin brought a complaint against Austin with Martin de Leon, the empresario of another colony. De Leon complained that Austin would not accept McLocklin's papers for the sale of a slave. In his letter, Austin named McLocklin as a member of a gang of "*picarros* on the Sabine" who sought entry to his colony, but "I ordered him to leave this jurisdiction without delay as the government had no use for such inhabitants."⁴⁹ Austin emphasized McLocklin's bad character by saying that "in the places where he was known it was sufficient only to pronounce his name to give an idea of all that is low and criminal in the character of man."

Four different people came to Austin to claim the slave in dispute, presenting papers he believed were obvious forgeries and falsehoods. This group of unwanted immigrants along the Sabine, evicted from Pecan Point and associated with McLocklin, included Nicholas Trammell and William English. Both Trammell and English were identified by Austin as men whom he "refused to receive in this colony, by infamy of their characters and which all the world proclaims by criminals and bad men."⁵⁰ From this account, two concepts emerge.

The first is that Nicholas Trammell was widely reputed to be one of the bad characters around the domain of early Texas, at least by Stephen F. Austin's standards. Austin also cast his net of criminality around William English in the same sentence. That statement was particularly interesting given that Austin breakfasted at English's house in the Ayish Bayou District back on his first full day in Texas in 1821.

The second and more interesting possibility that came from Austin's letter is the likelihood that Nicholas Trammell approached Austin in person, sought to become part of his colony, and was refused on the basis of his reputation. That Trammell and Austin met even earlier is a likely scenario. In 1819, Austin lived near Fulton, Arkansas, at a time when Trammell's trail was advertised as the most direct route into the colonies in Texas from Missouri and Tennessee.⁵¹ During that same time frame, Trammell and his kin were moving horses down the Trace to Nacogdoches, stirring up Indian relations, and expanding their unlicensed trade across the fluid national boundaries of the time. As a circuit judge in southwestern Arkansas, Austin may have already known about Trammell's reputation and activities even before coming to Texas. Whether or not Austin's refusal to admit Trammell into his Texas colony was a face-to-face rejection remains an unanswerable, but intriguing, question. In any case, Austin stated in no uncertain terms that Trammell was unwelcome in his well-governed colony. Trammell retreated toward Nacogdoches as a result of Austin's rejection, where he was soon caught in the middle of a clash over land that led to armed rebellion against Mexico.

1824, June—The Good Citizen of Mexico

Nicholas Trammell quickly became a fully involved resident of Mexican Texas. He pledged his allegiance to his new country on June 5, 1824, when he signed a statement saying he was willing to support the laws of Mexico.[52] It was an oath not taken seriously by many Anglos. Only days later, his pledge to uphold the law was acted on when he testified as a witness to a murder in the woods outside Nacogdoches.

On June 14, 1824, Trammell testified that he saw Aden Bunch murder Peter Young three days earlier. Like many others mentioned in the record with Nick, Bunch and Trammell had connections back up Trammel's Trace in Arkansas.[53] Trammell was briefly allied with Bunch in Nacogdoches in February 1824 when Trammell, Bunch, and William Boyce attested that a man named Romano was not guilty of a crime.[54] Back in Arkansas, Bunch was granted a license to operate a ferry across the Red River at Fulton in 1821 but abandoned or sold it not long after that to come to Texas. Bunch got rid of all his possessions in October 1821, collecting $517 from Elizabeth Griggs when he sold "all my house furniture containing 4 trunks, one bed and furniture with bedstead, kitchen furniture and several other articles, notes and accounts which I have against people in this county."[55]

Interestingly, Elizabeth Griggs soon became his wife, but it seems that the relationship began before Bunch was divorced. In the same month that he sold all of his assets to his future wife, he sued his current but absent wife, Bellona Bunch, for divorce in the Superior Court of the Territory of Arkansas.[56] By December 1821, Bunch was apparently on his way to Texas. He was summoned in Arkansas that month to answer a suit for payment of a debt of $60.50 to Joshua Morrison but was nowhere to be found.[57]

Based on Trammell's testimony, Peter Ellis Bean, as alcalde, ordered Bunch's arrest for murder. On June 17, Trammell gave more details of the crime he witnessed. Trammell saw Bunch ride into a camp where Young was present and then heard a "lick"—either a gunshot or a fatal blow. Trammell, from his hiding place, saw Bunch ride off in a hurry and followed him well out of view. When he drew closer to Bunch on horseback, Trammell saw him and a "Negro" driving a jack (mule) and two horses toward the Angelina River. Justice was swift, and on June 17, 1824, Aden Bunch was found guilty of murdering Peter Young.[58] No doubt the sentence was death by hanging. Elizabeth Griggs Bunch was a widow, and events soon led her husband's accuser to her doorstep.

1824, June—Race for Pig

June 1824 was a busy month for Nicholas Trammell. Serious matters like taking his oath of allegiance to Mexico and helping send a murderer to a death sentence were balanced with what started out to be a common Trammell diversion. One of Trammell's sons, Phillip, rode in an impromptu horse race. A dispute about the spoils of that race—a female pig—led to another court appearance for Nicholas Trammell.

The race was against a horse owned by a W. Homes (Holmes). Mr. Holmes put up the pig for his bet, but the record does not tell us what Trammell wagered on the race. The location for the race was not specified, but there were opportunities for a race either along an unprepared course or at one of several tracks in the Nacogdoches District. Samuel Steadham's racetrack was about twenty miles east of Nacogdoches on high ground along the Attoyac River. The race might also have taken place south of Nacogdoches at a place called Bodan, or even across level ground at Mustang Prairie near the Trinity River.[59]

Phillip Trammell won the race, two times in fact, but a dispute over the spoils made its way into the court in Nacogdoches two months after the race. Two witnesses gave testimony of the facts to Peter Ellis Bean. Cole and Fryley described the circumstances in detail, although Bean's poor English made it difficult to follow.

> Province of Texas, Naches, August the 7th, 1824.
> This Day Came John Cole and Solomon Fryley Before Me and maid oath that Some Time about the first of June Said Cole and fryley was at work in the field and that they caled out of the field to Juge a rase Between a boy of Nicholas Tramels by the name of Philip Tramil and a W. Homes and they Run the Rase and Tramil win the Rase for a gilt sow and after it was Run homes said that his Rider Road Jocky and it must be Run over again then homes Road him Self and they Run the Second Time and Said Tramil and him Run the Second time and Said trammel win the Second

Rase and then in the Presants of the two above Mentioned Witnesses Said Homes told trammel to take the gilt sow whare he could found her for Shee was his Property and but a few Days Sence Tramel found the Said Sow at the wido Bunches and Elison York clamed Said Sow and Tramel was thare and told him that it was His hog and forwarned york from taking of Said Sow York than ofred Tramel half of the hog But trammel Said if he did not get all of the hog that he would not have any of it and then Said york did not take the hog But after tramil was goan york took the said gilt sow.

A second statement on the matter provided testimony as to the possible whereabouts of the disputed pig: "The Declaration of John Foster, August the 12th 1824 I saw Elison york about the 20th of July take the Said gilt that Philip Tramel claimed and cary it off that he took it and Put it in the old house of John york."[60]

We can hope that York's old house was not inhabited at the time the pig was hidden there. The testimony from Foster after he saw the well-known pig, around July 20, is apparently what inspired Nick to peaceably bring the matter before the alcalde of Nacogdoches and attempt to resolve the dispute.

Bean's mention of the widow "Bunches" referred to the former Elizabeth Griggs, the wife of Aden Bunch, who was hanged as a result of the testimony of the man on her doorstep, Nicholas Trammell. Perhaps this was York's way of testing whether Trammell would face Bunch's wife in order to collect the sow. Perhaps York's presence at her house meant he had taken up with her after her husband's unfortunate end.

The lifetime interests of Nick and his sons in horse racing and their peaceable intent to collect their sow provide some insight into both the legal encounters Nick had and his interests in racing and gambling. The fate of the pig is not described for the record, but things only got worse for Nicholas Trammell as his time in Texas continued.[61]

1824, July—Slave Purchase from Jackson

Trammell's problems in Nacogdoches continued to mount as a result of the reputation that preceded his immigration from Arkansas. A business transaction—the purchase of a slave in 1824—led to eight years of courtroom troubles for Nicholas Trammell and his name again being brought before the eyes of Stephen F. Austin.

On July 7, 1824, John G. Jackson sold an unnamed slave to Nicholas Trammell for a note of debt in the amount of $890. Jackson also sold a couple of horses to other buyers around Nacogdoches. John Sprowl, alcalde for the Ayish Bayou east of Nacogdoches around present-day San Augustine, purchased one of the horses and later learned it was stolen. Sprowl also believed that the slave Jackson sold to Trammell was in fact a free man. Sprowl wrote Stephen F. Austin to warn him that Jackson, a man of questionable character, was on his way toward Austin's colony. "He is a large man, of genteel appearance about 25 years old, he came from near Alexandria on Red River, to my neighborhood, about two Months ago and brought with him two Horses and two Negros one horse he sold me, said horse I am told by undoubted authority the said Jackson must have s[t]ole from Esq Stokes in the Town of Rapede [Rapide, Louisiana]. Also one of the Negros is Stated to be a free man that Negro he has sold to Nicolas Trammal Near Nacogdoches."[62]

The transaction with Jackson came back to trouble Trammell. On May 10, 1825, Jackson sold Trammell's note of debt for $890 to John M. Bradley of Hempstead County, Arkansas. That transfer would surface again years later.[63]

A Bad Year in Texas

The unsettled nature of the boundaries and the unsettling presence of the people who migrated to the area around Nacogdoches drew the region into an unsustainable tension. Personalities and the forces of history converged there in 1825 and 1826 in ways that set another stone in the foundation of Texas' independence from Mexico. Not only were those years pivotal for Mexican Texas, but they were a turning point for Nicholas Trammell. He was not the type of person who chose to be in the middle of turmoil. He was not one who needed to be out in front of any activity or movement. Circumstances changed quickly for him. Nicholas Trammell was the focus of attention of local factions at odds over colonization contracts and both Anglo and Mexican authorities in Texas. As a result of Trammell's entanglement in events with historic ramifications, he and his family were forced into a complete reversal of their good fortune in Texas.

During those years in Nacogdoches, old settlers and new immigrants around Nacogdoches were wrangling for control over thousands of acres of land. Overlapping colonization contracts were given to empresarios with competing interests. Those who lived around Nacogdoches for many years and received Spanish land grants were suddenly in competition for land with new immigrants. The reputations of the Anglos who sought control were built on the cleverness of their land schemes or the extent to which deviousness and bloodshed might be used to carry them out. Unlike many, Trammell appears to have had no interest in power or notoriety. It simply was not his style. His involvement with people who did have those interests led him into a series of events that ended with his ignoble return to Arkansas. After Trammell's legal difficulties in 1824, one short week in early 1825 provided a snapshot of both his best opportunity and his miserable collapse.[64]

1825, March—Slave Gabriel, Rifle, and Horse Sold

On March 24, 1825, the Mexican government issued a Colonization Act for the State of Coahuila and Texas. One of the early settlers during this year of monumental change put the Colonization Act of 1825 in perspective by later saying, "the discovery of gold was to California what the colonization act of 1825 was to Texas."[65] The act encouraged immigration and specified how it took place outside the structure of the earlier colonization contracts. By pledging an oath to support the government and promising to live the life of a faithful Catholic and Mexican subject, virtually anyone could receive hundreds of acres of land for a very small price.[66]

Only one week before the act that provided this great opportunity went into effect, Nicholas Trammell began a well-documented descent to a low point in his life when the two most prized possessions of a man of the frontier—his rifle and his horse—were sold at public auction.[67]

Attorney Leonard Dubois filed a claim against Nicholas Trammell in Nacogdoches on behalf of Francois Villon. Villon's attorney reported that a slave named Gabriel, who was her property, was in Trammell's possession.[68] Testimony on the matter was taken in Nacogdoches, but Dubois also sought a judgment from Stephen F. Austin in the District of San Felipe de Austin. Even though the case was filed in March, the claim stated that Trammell stole the slave over six months earlier, in September 1824. It is unclear why it took so long for Villon to seek redress.

Trammell's poor reputation beyond the Nacogdoches District added drama to the matter. John P. Coles, one of Austin's original Old Three Hundred settlers, wrote Austin on May 13, 1825, informing him that Gabriel could be at one of several locations and that an officer should accompany Dubois to look for the slave.[69] Dubois must have been concerned for his own safety since he sought an escort. Coles also acknowledged, "they are all concerned with Tramel."[70] William Pryor, another Austin colonist, gave testimony to Austin regarding the slave.[71] Pryor told Austin that in January 1825, Nicholas Trammell visited Pryor at his house on the Brazos River. Trammell told Pryor that he and a woman from Nacogdoches were in dispute about possession of the slave and that Gabriel was at Trammell's house.[72]

Daniel Quinn testified that he saw Gabriel at Trammell's home at the Angelina River crossing as early as November 1824.[73] Gabriel spoke with Quinn but would not tell Quinn his real name, instead saying that his name was John. Quinn recognized him as Gabriel, however, and realized that he either had been stolen by Trammell or had run away from Dubois. Quinn's testimony raises an interesting question about the nature of the transaction and perhaps some insight into Trammell's treatment of slaves: "the said negro told him [Quinn] that Duboy had him chained that Tramel came and cut him loos and afterward said Tramel said there was too much truth in what the negro said and therefore this deponent sayeth not."[74]

Quinn's testimony, that Dubois had Gabriel chained and that Trammell came to free him, raises the possibility that Trammell did so either out of compassion for Gabriel or to regain property he believed to be his own. In either case, Gabriel appeared to be in no hurry to leave Trammell's service or his protection from Dubois's chains.

After the legal arguments, Samuel Norris found in favor of Villon. Trammell apparently refused to pay the debt or return the slave. Instead, two of Trammell's key possessions were auctioned. Trammell's horse was sold to Joseph Durst for fourteen dollars and his rifle to Leonard Dubois for nine dollars. How the court took possession of the property is unclear, since these two items were among the last a man would willingly surrender. Equally unclear was the reason that any of Trammell's other property was not sold at auction and that twenty-three dollars satisfied the debt. Nevertheless, it was probably not a happy occasion.[75] Nicholas Trammell's sense of frontier fairness was not something to be taken lightly.

1825, August—Slave Sophia, Andrew Hemphill

Besides the lawsuit over a slave sale in Texas, another unresolved transaction involving a slave in Arkansas resurfaced in July 1825. Prior to leaving for Texas, Trammell sued Andrew Hemphill, a member of a prominent family of salt makers. Trammell purchased a slave named Sophia from Hemphill for thirty-four dollars and two horses valued at sixty-five dollars. Sophia was described as ten years old, yellow of complexion, and a slave for life. Hemphill gave Nick a bill of sale, held in the office of Thomas Drew, clerk of the court in Clark County, Arkansas. Trammell testified that he could not write his name or "letters with a pen," but the note supposedly showed Trammell's signature. Even though Trammel initiated the legal action, the judge in Arkansas held a hearing in Trammell's absence that resulted in a judgment against him.

A friend of the family, Jacob Barkman, learned of the hearing, represented Trammell in court in his absence, and immediately appealed the decision on his behalf. As Trammell's approach to collecting his race winnings demonstrated, he could be patient but did not take kindly to being treated unfairly. In July 1825, as part of the appeal process, Zachariah Davis and Jacob Barkman came to the "Spanish Country and did prove a Certain Negro Girl named Sophia was in possession of Nicholas Tramel." Barkman testified that Trammell proved Sophia to be his property and took her lawfully, in spite of the earlier ruling.

Nick countersued after the judgment was made against him. Believing justice was misapplied, he also included the justice of the peace and the county clerk as defendants in his suit.[76] Trammell's troubles with this case would continue for almost seven years.

Trammell at the Trinity

Before a storm, the winds blow toward the center of low pressure behind the front. As hard as the wind blows before the storm, it will blow equally hard in the opposite direction when the storm arrives. The winds of change blew favorably on the Trammell family after his rejection by Stephen F. Austin. When Trammell was not accepted into Austin's colony, he and his family retreated toward Nacogdoches. There, he was able to reestablish himself in a position that had many advantages for his business in contraband and horses. Nicholas Trammell and his family settled at the busy Trinity River crossing of the El Camino Real.

After Trammell moved to eastern Texas, Mexico gave a colonization contract to settle up to eight hundred families in the area around Nacogdoches to a land speculator named Haden Edwards. Edwards arrived in Nacogdoches on September 25, 1825, prepared to force his influence in ways that the Mexican authorities did not anticipate. The law required that Edwards honor all grants previously made by Spain and Mexico. His tactics for meeting that requirement infuriated the settlers. Haden Edwards and his brother, Benjamin, required all prior landowners to prove their claims or forfeit their land and have it assigned to someone else. The burden of proof was on people who had been on their land for years but who could not always produce Spanish documents. Tensions around Nacogdoches became intense as a result of conflict over land and power between the old settlers and the new Anglo colonizer. That conflict resulted in a short but notable rebellion in which Nicholas Trammell was the spark.

Within only two months of Haden Edwards's arrival in Nacogdoches, he sold one league of land at the crossing to Nicholas Trammell's eighteen-year-old son, Nathaniel, a transaction that drew the attention of many. A $120 down payment on November 25, 1825, began Trammell's claim to a league of land, half on either side of the Trinity River, which included the key ferry crossing on the El Camino Real to San Antonio.[77] Edwards nodded toward Mexican law by asking Nathaniel for a certificate of good character. James Tate, a former judge in Louisiana, offered his former title and certification of good character to the scheme by Trammell and Edwards, but little else. Tate had his own questionable character to defend after abandoning his nine-year-old stepson outside San Augustine in 1821.[78]

Surveys along the Trinity typically offered one-half league of river frontage and stretched two leagues away from the river and all on one side of the river or the other. The surveyor in this instance specifically used the ferry crossing as the center of the survey, giving Trammell control of both sides of the river. This put Trammell in an extremely favorable position for ferrying soldiers and travelers—a location where he could easily participate in the transport of goods to and from Nacogdoches.

Speculation about what led to Trammell's settlement at the Trinity River is supported by little evidence. In these times, Nacogdoches was a bit of a gambler's heaven, but it was quite far to the east of the Trinity. Perhaps Haden Edwards believed that having the allegiance of someone of Trammell's character at a key crossing helped secure his western flank from a surprise infusion of Mexican soldiers. Trammell's reputation for loyalty, when suitably compensated by familial ties or money, may have led Edwards to make such key property available. Perhaps they shared a common interest in gambling. Edwards reportedly operated a roulette table in Mexico before he came to make trouble in Mexican Texas. There is some indication that Trammell had settled along the road near there already and merely took advantage of Edwards's arrival to secure a questionable ownership of land. Trammell's presence at that critical crossing was documented as early as

November 1825, but he may have been living there since the year before. Daniel O'Quinn testified that he saw a slave at "Nicholas Trammels on Trinida River" in November 1825.[79]

This early evidence of Trammell's location is interesting for several reasons. First, that Trammell was located well to the west of the political turmoil in Nacogdoches, the only settlement in the eastern part of the Mexican territory. From Nacogdoches, it was twenty miles west to the Angelina River crossing of the El Camino Real, two miles past John Durst's. After crossing the Angelina, Peter Ellis Bean's property was eight more miles, only a couple of miles from the Neches River crossing. Trammell's home on the Trinity River was ninety miles, three to five days' journey, west of Nacogdoches.[80] That placement reflected a common trait throughout Trammell's life. He was often found somewhat away from major settlements, out of the direct gaze of any legal authority that might inhibit his trade and gambling activities.

Second, that he was drawn to an intersection of a key waterway and a well-traveled roadway was unsurprising. This critical crossing near the location of old Trinidad de Salcedo was proximal to a place where people and goods slowed to cross the river and the wet floodplains on either side. Trammell was not only able to take advantage of the opportunity for trade at a natural rest stop, but he was in a place that allowed him to use one of the clandestine roadways that bypassed Nacogdoches for moving horses from the interior to the United States. The commercial traffic leading past Trammell's door provided ample business opportunity for him and his kin.

The problem with the sale was that the land Edwards conveyed to Nathaniel Trammell was already settled by Ignacio Sartuche.[81] Testimony varied, but it appears that Trammell initially lived there cooperatively with the old Mexican settler, at least for a period of time. Their residences were both close to the river crossing, and Trammell supplied the poorer Sartuche with food and supplies. Sartuche claimed to have lived at the Trinity River crossing since 1821 or 1822, before Trammell's arrival on the scene, where he built a home near Loma del Toro (Bull's Hill), planted crops, and hoped to maintain his family.

After the Long invasion of Nacogdoches in 1819, Mexicans and older settlers in the area were suspicious, even hostile, toward Anglos, particularly any who tried to rule over them with their own laws. Edwards's favoritism toward the Trammells quickly led to a broad dispute when the transaction involving Sartuche became public. One of the aspects of the colonization law that came to weigh heavily on Nicholas Trammell was that in the distribution of lands by the empresarios, preference was to be given to Mexican citizens. From the law's perspective, the old settlers were to be given the benefit of the doubt. According to Edwards, however, if there were no documents, there would be no ownership, no matter how long any old settler may have occupied his land.

Sartuche offered evidence in December 1824 to the chief of the Department of Coahuila and Texas, Don Jose Antonio Saucedo, that his claim was valid and predated Edwards' award of the colonization contract. In spite of the apparent legitimacy of Sartuche's claim, Edwards sold Sartuche's land to Trammell. Sartuche petitioned the court to regain his land, and Saucedo acted promptly on Sartuche's request to be restored to his property.[82] Saucedo delivered the petition to Samuel Norris with an order to restore Sartuche to the land and remove the foreigner Trammell. Nacogdoches's magistrate, Luis Procela, also supported Sartuche's claim and pointed out the numerous transgressions of the new empresario, Edwards, including his confiscation of the archives and his attempt to steal an election for alcalde. Inciting emotion by stating that the citizens of the Mexican Republic would become slaves if they allowed such behavior, Procela observed the audacity of Edwards and Trammell in taking the land: "It has been but a few days since the Empresario went to the Trinity River and gave to a certain Tramel the land of the Citizen, Ignacio Sartuche, notwithstanding the decree of Don Gaspar Flores [which said that Mexican land claimants should be preferred to foreigners], and thus the Empresario will continue in this manner, taking and disposing of lands belonging to our citizens."[83]

The Trammells continued their semblance of legitimacy by making payment on March 3, 1826, for the survey of the land granted them by Edwards. Thirty dollars could not legitimize the transaction.[84]

Samuel Norris, the alcalde for the old settlers, was having none of this. Edwards was a latecomer to the region, granted a colonization contract without due diligence on the part of the Mexican government, and was viewed by the old settlers as an illegitimate claimant to the responsibilities as empresario. It was not just Edwards's behavior that created confusion. His empresario contract was also questioned. Peter Ellis Bean was also awarded an ill-conceived contract that overlapped with lands twenty leagues from the border with the United States. The boundary reserve was land within twenty leagues of the border protected by the Mexican government from any Anglo settlement. The boundaries for Bean's contract conflicted and overlapped with those outlined for Edwards. In a letter to Stephen F. Austin, Benjamin Edwards complained about the growing problems: "It is reported and believed that Been [Bean] has a grant to the Nuteral lands; and yet the Alcalda [Samuel Norris] is letting out lands to his favourites; suffering them to take the improvements of others, while some rely upon obtaining their titles through Been or his commissioner, and all doubtful of the security of their lands eventually."[85]

Samuel Norris was prepared to take decisive action against Trammell and Edwards by proxy. In March 1826, Austin wrote a blistering indictment of Trammell and his "band of picarros." Emboldened by that action, Norris gave the order in April to arrest Trammell for stealing the slave Gabriel.[86] Norris did everything in his power to dislodge the Trammells and make them feel unwelcome in Texas, but there is no indication that Trammell was ever arrested or taken into custody in Nacogdoches. Norris's animosity toward Edwards was ultimately affirmed on August 23, 1826, when the Mexican government annulled Edwards's empresario contract. With the support of the old settlers and all of the available legal authority on his side, the emboldened Norris was about to deal a final blow to this land deal and end Nicholas Trammell's presence west of Nacogdoches.

Disputes over land were not the only source of conflict. Increasing pressures from migration, shifting alliances among the Anglos, Indians, and Mexicans, and various authorities who claimed rights to land or hunting grounds led to a volatile mixture of people and events. While Trammell was apparently away from home, Indians crossing the Trinity to potentially threaten the eastern settlements visited Trammell's wife, Sarah, at their home near the Trinity ferry. A man named Foster, whom Austin sent to confer with the Cherokees, spoke with Sarah about her encounter. The Trammell family's goodwill with the Indians may have saved her life: "at the Trinity he [Foster] was told by Mrs. Trammel that 5 Keechis had been there the day before and that they told a spaniard that they intended to kill all the Americans. She took an old Aynye [Avaye] chief into the house and asked him and he told her yes it was true they were all mad but would not kill her as she had treated them well."[87]

The location of Trammell's land at the Trinity was a pinch point of commerce. Just east of Trammell's ferry crossing, a road left the El Camino Real to the north and led to a strategic resource—a saline, or salt lick, that had been used by the local Indians for many years. Peter Ellis Bean lived along the road and at one time owned the saline, harvesting salt in commercial quantities. A road from Bean's saline toward Pecan Point provided a cutoff toward Trammel's Trace to the north and a means by which Trammell could

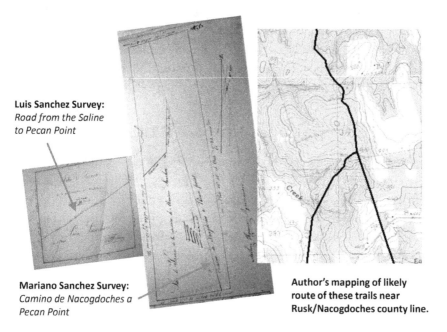

Roads to Pecan Point from Nacogdoches in Mariano Sanchez and Luis Sanchez Land Grants. Plats from East Texas Research Center, mapping by author.

move goods between the Trinity River and the smugglers' road named for him without going through Nacogdoches.

Travelers from the interior of Texas toward the United States, no matter what route they took, converged at the Trinity River ferry. All roads led to Nicholas Trammell.

An Eviction and a Killing

After considering the facts and the pressures to act, Samuel Norris ordered Nicholas Trammell to vacate the land at the Trinity River crossing of the El Camino Real on September 29, 1826. Trammell, his children, his wife, and various others, including his half brother and business partner, Mote Askey, had to go. Norris did not care where they went—he simply and bluntly told them to leave: "Mr. Trammel, you are hereby ordered to quit the plantation of Mr. Sartuches without any molestation to him or anything that is his. If you are not out in fourteen days from this date, I will send the militia and put you off at your own expense."[88]

The clock was ticking, and Trammell was out of options. Before the order could be acted on, an even more unexpected incident occurred—Mote Askey was shot and killed two weeks later. Trammell's half brother was not killed by Indians, nor was he the victim of a shady transaction gone wrong. He was killed by a future signer of the Texas Declaration of Independence.

1826, October 13—Mote Is Killed

Mote Askey, his brothers Otho (Otto) and Abner, and his father, Zachariah, were closely allied with Nicholas Trammell throughout his life.[89] Mote was married, and his wife and children were in the Nacogdoches District as well. Besides the Cherokee petition in 1813, Mote appeared in census records along the White River with Trammell as they migrated together from Kentucky to Missouri, Arkansas, and Texas.[90] Mote Askey was shot on October 10, 1826, and died three days later, just in advance of Trammell's deadline for leaving the ferry crossing. Mote's brother, Otho, gave his statement of the events to Samuel Norris: Otho "has Just Cause to Believe Martin Parmer Did Shoot Moton Askey near the house of Daniel Clark on the Anneline (Angelina River) and that said Moton Askey Died on the 13th of the same month of said wound."[91]

Martin Parmer was a relative newcomer to Texas. He arrived in 1825 and initially lived just east of the Angelina but moved to the Ayish Bayou region between Nacogdoches and the Sabine, part of the Neutral Ground, in early 1826. Parmer was known in Missouri as the "Ring-tailed Panther" and was described as uneducated, unpolished, aggressive, and daring. In his earlier homes in Tennessee, Kentucky, and Missouri, he served as a leader in local militias and was active in battles with Indians.[92] Alcalde Norris issued an order for the sheriff to take Parmer into custody. There is no evidence to suggest Parmer was ever called to answer for Askey's death. Any attempt to arrest Parmer was quickly sidetracked by a rapid chain of events.

The more interesting questions are about the possible reasons for the shooting. With Mote's death so close to the deadline for Trammell to vacate the Trinity crossing, the Trammell clan must have been in an uproar trying to make decisions about what to do. Parmer was part of a growing number of men who were armed and prepared to incite the Indians against Samuel Norris and the Mexican government. Perhaps Parmer was recruiting Nicholas Trammell and Mote Askey to join the rebellion and Askey declined. Nicholas Trammell's present difficulties put him in no position to request justice from Norris, but there can be little doubt that he was indignant and incensed about Mote's death. To add even more burden to the Trammells and more sorrow to the Askeys, Mote Askey's wife was expecting a child at the time he was killed. There was little time for mourning, much less for justice.

1826, October 20—Trammell Chased out of Texas

The day Mote died was the deadline for the Trammells to leave. Nicholas Trammell's reckoning at the Trinity crossing came seven days later, perhaps to allow the family time for burial and mourning. That time quickly ended, and action against Trammell continued. Acting on Norris's order, an armed militia rode out to arrest Trammell on October 20, 1826. Two unnamed men were arrested, but they later escaped across the Sabine. Nicholas Trammell and "other bad men" escaped capture and reportedly fled to Pecan Point. There was apparently no pursuit.[93] Trammell's land and livestock were left behind in Texas, and his family was scattered. Unless Trammell made prior plans for his property, there was no opportunity to remove possessions from his house and no information about the disposition of any slaves he owned. There is no way to know if he was expecting the attack and prepared supplies or made plans or if he was forced to leave with only what he could carry on a horse.

Six days after Trammell's eviction, Ignacio Sartuche's Trinity River land claim was resurveyed. Sartuche walked the boundaries with the surveyor and exercised the typical Spanish demonstrations of ownership. Sartuche gave thanks to God as he pulled up weeds and tossed them in the air. At each of the corners, he tossed stones and cried aloud, perhaps tossing one in the direction of Trammell's hasty retreat. Dirt and stones flew behind Trammell as well, but only those loosened by his horse's hooves as he escaped north up Trammel's Trace.

7

1826–1836
A Great Movement of Many Nations

Either I am mistaken, or this is a development of the greatest importance to the history of the human species.

—MANUEL DE MIER Y TERAN[1]

In the first quarter of the nineteenth century, history carved two paths across Arkansas and Texas, one literal and one figurative. The literal path was Trammel's Trace, the former smuggler's trail that provided two of the three points of entry into Texas from the United States. The figurative path was that taken by Nicholas Trammell through the history of Texas and Arkansas. For almost fifteen years, these two paths were the same—Nicholas Trammell and the road named for him had the same purpose. Both the road and the man were about unlicensed and illegal trade in a foreign country.

Anglo settlers poured into Texas in 1821 and again after 1825, challenging immigration policies, land ownership, and governance. Nicholas Trammell came to Texas during that wave with many of the same hopes as other immigrants. He sought land ownership and a fresh opportunity for trade. Although he wound up in the Nacogdoches District as a result of his rejection for entry to the colony organized by Stephen F. Austin, Trammell's subsequent residence at a strategic road and ferry crossing on the Trinity River had the potential to become the pinnacle of his family's trading business. Trammell was in his mid-forties when he settled at the

Trinity crossing, his sons were becoming adults, and life must have seemed more certain and promising. What better place for a trader and a trailblazer to inhabit than a key crossroad on the fringes of a rapidly growing frontier?

In October 1826, Trammell was chased out of Texas and those two histories began to diverge, each taking a course independent of the other. Trammel's Trace directed settlers and immigrants to a different Texas in the years following Nicholas Trammell's rapid departure. Nicholas Trammell retreated to Arkansas and developed an altered approach to the family business.

Undoubtedly, Trammell was disappointed and perhaps angry when he was run out of Texas, but that was not reflected in any recorded acts of vengeance. Stories about borderlands justice detail many examples of killings or other acts of revenge. If any man had cause to seek retribution, it was Nicholas Trammell. His name was broadcast across the colonies as a horse thief and slave trader. Stephen F. Austin rejected him from acceptance into his colony. Trammel's half brother was killed and no one was arrested. He and his family were chased out of Texas by armed soldiers. The legal system had not served Trammell with justice by his standards, not even in the simple recovery of a pig he won as the prize of a horse race. Nevertheless, there is nothing to indicate Nicholas Trammell was the type of man inclined to gather his own armed band to return and seek retribution. Trammell's retreat northward on Trammel's Trace to Arkansas was back home to more familiar circumstances and to people he had trusted. When he was chased out of Texas, the career of Nicholas Trammell as a "bad man" of the Texas frontier was over. However, that just meant he had to regroup and enter more "legitimate" businesses like horse racing and gambling.

Repercussions in Nacogdoches

After Nicholas Trammell unwillingly relinquished his residence at the Trinity River ferry west of Nacogdoches, the land dispute that led to his expulsion was a cause célèbre for armed rebellion. Events in Nacogdoches sent a signal of growing discontent across Texas only one month after Trammell was run out of Texas. Even though the governor annulled his colonization contract, Haden Edwards stayed behind in Nacogdoches, but not to make amends. Benjamin and Haden Edwards planned to overthrow the Mexican government in Nacogdoches.

On November 22, 1826, thirty-six armed men entered Nacogdoches. Edward's armed action against Mexican authority, called the Fredonian

Rebellion, was a precursor to Texas' war for independence from Mexico almost ten years later. The filibusters arrested the commander of the militia, José Antonio Sepulveda, and the alcalde, Samuel Norris. They also offered a reward for Norris's brother-in-law, James Gaines, dead or alive. The leader of the rebellion against Mexican authority was none other than Martin Parmer, the man accused of killing Trammel's half brother, Mote Askey. Parmer and his men confiscated the Nacogdoches court records and archives and then installed their own alcalde, Joseph Durst. Three days later, Parmer presided over a court martial of Norris and the military commandant, but they were released after a brief show trial. The leaders of the uprising also arrested Haden Edwards, but only to maintain the appearance that the group's hollow claims of injustice were not connected to the former empresario.[2]

After a few days' occupation of Nacogdoches, the rebels left to regroup and assess their accomplishments. News of the revolt traveled quickly to Stephen F. Austin, and he offered his explanation of the matter in a letter to Governor Saucedo dated December 4, 1826: "From what I could learn of that occurrence, it would seem that the principal cause was the hatred of those people toward Gaines and Norris, and not any ill feeling against the Government. With an intelligent and impartial man to administer justice in that locality, no difficulty need be apprehended on the part of the inhabitants. There are, however, some bad and rebellious men, who must be expelled from the country."[3]

Austin underestimated the source of the rebellion, for the time being. Of even greater concern to Austin and the Mexican government than armed men ousting local leaders was that the Fredonians sought a pact with local Indians to unite in their battle against Mexico. The fear of Indians taking sides and joining in battle was a concern shared by both the Mexican officials and Anglo settlers. In exchange for their support, the Fredonian leaders promised the Cherokees their own land—that which they longed for most.

Haden Edwards and his brother, Benjamin, formalized their uprising when they issued the Fredonian Declaration of Independence on December 21, 1826. Austin was forced to reassess his earlier dismissal of the events as a minor dissatisfaction with local government. The threat of Indian war against the Mexican authorities, and an upset to his carefully planned and patient strategies for the colonization of Texas, led Austin to issue a call to arms. Austin broadcast his concerns all across Texas in January 1827: "A small party of infatuated madmen of Nacogdoches have declared Independence and invited the Indians from Sabine to Rio Grande to join them, and wage a war of murder, plunder, and desolation on the innocent inhabitants of the frontier."[4]

Some of Austin's old enemies were part of that "party of infatuated madmen," as were well-known outlaws of the region. Along with Martin Parmer, William English again took sides opposing Austin. The Colliers (Calliers) and the Yoakums (Yokums), two families guilty of crimes all along the border, were well-armed supporters of the rebellion.[5] Stories of killing and torture connected to the Yoakums led many blacks, both free and slave, to fear them. "A panicky terror seizes them as soon as someone utters his cursed name" is how one observer described the impact of these "brigands of the Sabine."[6]

The Edwards brothers negotiated an agreement for support by the Cherokee Indians living north of Nacogdoches with Richard Fields and John Dunn Hunter, two of the local chiefs. In late 1826, just before the Fredonian Rebellion began, tribes north of the Red River near Pecan Point suddenly moved to the south side of the river. Ben Milam represented the Red River region at that time as Indian agent for the Mexican government. When Milam arrived at Pecan Point in November 1826, he found the Anglos very upset about the organized southward movement of the Indians. Milam later discovered the stimulus for the movement—Chief Richard Fields called the tribes to come to Texas and implied that he had the power to distribute land, an affirmation of Edwards's plan. The Cherokee ultimately discovered the weakness of the Fredonian's scheme and withdrew support for their cause, support that was an ongoing disagreement within the tribal council.

In spite of that action, Mexican authorities were still convinced that the Cherokee would take arms against them based on an intercepted communication. With the future of the tribe put at risk by Fields and Hunter's action, Chief Bowles reasserted his leadership and took harsh measures in hopes of gaining favor with Mexico. Bowles ordered Fields and Hunter killed in retaliation for their digression. The lives of these two Cherokee leaders were expendable in the cause for Cherokee land, and pleasing Mexico was far more likely to result in land than supporting the Fredonians. The Cherokee would feel the effects of this chain of events in dramatic fashion. Thirteen years later, after Texas became a Republic, their very presence in Texas faced the ultimate challenge of expulsion.

When the Fredonian cause collapsed in defeat, Parmer and English fled east to the United States. They both returned to Texas years later and were part of the Texian Consultations in 1835, precursors to a declaration of

Texas independence. Out of necessity, these "infatuated madmen" later became allies of Stephen F. Austin.

Trammell Back in Arkansas
October 1826—Texas a Home No More

The doors to Texas closed behind Nicholas Trammell when he was chased out of Nacogdoches. When he retreated to Arkansas to regroup, he withdrew to the familiar. Nick and his family traveled back up Trammel's Trace along trails he had ridden many times, up the Southwest Trail and across hundreds of miles of territory through difficult conditions. In October 1826, Trammell returned to the White River country in northeastern Arkansas, which he left four years earlier, the place where he had originally settled in the Arkansas Territory.

Trammell may have understood when he left Arkansas for Texas that his possibilities there were not certain. Before he left, he kept his options open back along the White River. When Trammell left northeastern Arkansas for Texas in 1822, he sold his 640 acres to Morgan Magness. Although Trammell sold the pre-emption rights to Magness, Trammell still had a possible claim on the land through a bounty warrant, a claim issued as a result of his father's military service.[7]

When Trammell returned in 1826, Magness sought to exercise his influence when he petitioned the US Congress to "pass a law authorizing him [Magness] to surrender his right to the quarter section called for by the certificate, and select another quarter in lieu of it." Magness's petition in December 1826 followed Trammell's expulsion from Texas by only two months, so Trammell wasted little time in relocating and calling on old friends. In his petition, Magness offered an affidavit from Charles Kelly that supported Trammell's prior claim to the land. Charles Kelly, the sheriff who testified to Trammell's initial claim to the land, said he was "well convinced that Trammell intended to obtain a certificate for this quarter section of bounty land from the circumstance of Trammell's being settled upon it." Kelly seemed to imply that Trammell was simply away on business and returned to his land as planned. Curiously, it was Magness who offered up this information. Magness's approach seems to imply that he perhaps knew about Trammell's second claim to the land or at least was not upset in giving it back to Trammell. Perhaps Magness and Trammell envisioned an easy congressional approval of Magness's request, given his wealth and prominence. A positive outcome for each of them seemed possible if only

Congress had accepted the proposal. Trammell would have gotten his land back and Magness would have received another claim. Congress denied Magness's petition by saying "the objections to granting this indulgence are too obvious to require enumeration." Nicholas Trammell was back home on the White River, but the ruling of the House left his status unsettled, and that would never do for Trammell.[8]

1827–1828—Canebrakes of Lost Prairie

Trammell's return to the White River was short lived. Although he sought the comfort of his old territory and trading partners to recover from his treatment in Texas, he did not linger there. The White River country was far from the frontier where business prospects loomed. If nothing else, Trammell's involvement with the primary actors in the Fredonian Rebellion convinced him that Texas independence from Mexico was coming sooner or later and that immigration, and therefore trading opportunities, would continue to rapidly increase. Trammell and his family wasted little time in returning closer to the boundary of the United States and Mexico, nearer to Texas. In 1827 or 1828, Trammell relocated to Lost Prairie, Arkansas, in Lafayette County, south of Fulton in the flatlands between the Red River and Mexican Texas.[9]

The fertile soil in the alluvial plain was covered by thousands of acres of tall, thick cane. Years of flooding along the Red River left rich deposits behind. Seven miles of canebrake lined the west side of the Red River at Trammel's Trace across from Fulton. Deeper in Texas there were accounts of a creek-side canebrake seventy miles long. Anyone who wanted to pass through such thick stands of cane faced a unique obstacle, overcome only by the regular crossing of others gone before: "The frequent passage of men and horses keeps open a narrow path, not wide enough for two mustangs to pass with convenience. The reeds grow to the height of about twenty feet and are so slender, that having no support directly over the path, they drop a little inward, and so meet and intermingle their tops, forming a complete covering overhead... and with the view of the sky shut out."[10]

Bear hunters ventured into the denseness of Lost Prairie at times, but others had no reason. Within those shadowy confines, Nicholas Trammell and his family sought refuge. Trammell's seclusion only added to his mystery and legend. It was said that his farm was "surrounded on all sides by almost impenetrable canebrakes, but no road leading to and from it had yet been discovered."[11] Trammell's mysterious occupation and growing reputation led others to grant him some magical power to enter his hidden enclave without the benefit of a trail.

That description of the landscape was supported by a surveyor's observations some years before. An 1823 survey of the area showed the only cleared piece of land in the entire township was a single field at the foot of the bluff near a ferry crossing.[12] Anyone who crossed the Red River at Lost Prairie was faced with three or four miles of thick, muddy bottom land. The only elevated places were a few slight hills and old Indian mounds. Beyond that were two thousand acres of incredible beauty and fertility. After it was cleared of the cane, the flat, fertile land between the Red River and McKinney Bayou to the west became prime ground for cotton growers with large plantations and significant holdings of slaves.

True to their previous histories, many of the early settlers who came to the Lost Prairie Township of Lafayette County, Arkansas, were those with years of connections with Trammell. John Barkman, Andrew Hemphill, Collin McKinney, Joshua Morrison, and Trammell's oldest son, Nathaniel, were all on the tax rolls by 1828. Ferry operators James Byrnside, George Dooley, and Bryan T. Noland (Nowlin) were all there as well. Two of Trammell's other sons, Phillip and Henry, also had property on the rich land along the road from Dooley's Ferry heading toward Trammel's Trace.

Nathaniel Trammell also owned land at a key site for trade in Chicot County in far southeastern Arkansas at the Mississippi River. An unclaimed letter for Nathaniel waited in the Chicot County post office in December 1828. Although his father settled in Lost Prairie that year, Nathaniel Trammell did not appear on the tax rolls in Lafayette County until 1829. Records indicate Nathaniel was established in both Chicot County and Lafayette County during the same year.[13]

Just prior to his move to Lost Prairie, Nathaniel Trammell also appeared in the records back in Nacogdoches. Given all the controversy over the ownership of the land at the Trinity River on the El Camino Real, that Nathaniel Trammell was able to retain ownership is surprising and unclear. On October 28, 1828, a court record shows that Nathaniel Trammell sold the land at the Trinity ferry crossing to James Tate, the man of questionable moral fiber who signed the character certificate supporting Nathaniel's initial purchase.

With their relocation near key river crossings all across the entire width of southern Arkansas, the Trammells were putting a business plan in place. If Trammell had learned anything about the turmoil he left behind in Texas, it was that men had designs for Mexican Texas. The schemes of men focused on Texas were a response to the coming movement of many people and much trade.

1828—Bradley Debt

Although his business prospects seemed to be in a position to improve after he returned to Arkansas, Trammell's legal affairs in Arkansas remained unsettled. A history of lawsuits involving debt remained a constant in his life, dragging on for years after his return to Arkansas. Trammell's promissory note of $890 to John G. Jackson for the purchase of an alleged free man of color, which had surfaced while he was on the Trinity River, resurfaced when Trammell moved to Lost Prairie.

Jackson accepted the note from Trammell but later sold it to rid himself of the note, as was common in the economy of the day. Jackson sold Trammell's bond of debt on May 10, 1825, to John M. Bradley. Bradley was another early resident of Lost Prairie in Lafayette County, Arkansas, near Trammell's new home.[14] Bradley claimed that he repeatedly notified Trammell that the debt was due, starting in November 1825, when Trammell was settled at the Trinity crossing west of Nacogdoches. Bradley's notification may have only been in the form of a lawsuit filed in Arkansas or a notice published in the newspaper because Trammell was in Texas and Bradley was still in Arkansas. There is no indication that Trammell learned of the suit until he settled in Lafayette County, where Bradley lived. Trammell may have considered the debt long gone since there was no action for three years after the note was executed. After Trammell moved to Lost Prairie, the judge in Lafayette County instructed Joshua Morrison, the first sheriff of the county, to take Trammell into custody in July 1828 and require a bond of $1,000.[15] Trammell posted bond with the help of his son, Nathaniel, and the former Sulphur Factory trader Robert B. Fowler.

Trammell wasted no time in defending himself against Bradley's claim, retaining Chester Ashley as his attorney. Ashley, the premier appellate attorney in the Arkansas Territory, was associated with the James Sevier Conway political faction from Lost Prairie, which dominated early state politics. Conway was later a governor of Arkansas. Earlier that same year, Ashley reportedly shot a deputy sheriff from Pulaski County, Arkansas, but was not convicted.[16] Ashley was a land speculator himself and also represented James Bowie in his many fraudulent land claims. Ashley was just the kind of attorney one would expect Nicholas Trammell to retain—capable, connected, and able to dance along the boundaries of legality.

Ashley filed a motion with Judge William Trimble to dismiss the suit on July 22, 1828. The claim was continued to the August 1829 term of the court and then delayed again to April 1830. By that time, another magistrate, Judge Benjamin Johnson, was on the bench, and Trammell filed yet another request to dismiss. A third judge, Thomas P. Eskridge, assumed the case and finally ruled on October 19, 1830, that Trammell pay Bradley the $890 shown on the note. Trammell and his attorney filed an appeal the very next day, but only on a point of law, not on the question of the debt. A significant fact came to light only after the judge's ruling. Bradley lost the note and could not produce it for the court. Since the note could not be produced, the crafty Ashley simply argued that Bradley had no right to make the claim in the first place.

Further delays came when Judge Eskridge permitted the parties to seek depositions in the matter. Bradley sought approval in February 1831 to locate witnesses in Louisiana and Mississippi. The case was delayed so long that it appeared before a fourth Circuit Court judge in April 1831. Judge Edward Cross was former counsel to Bradley on this same case and therefore could not make a ruling, so Cross sent the case to the Superior Court for the Arkansas Territory, the fifth court to hear the matter. Finally, at their January session in 1832, the Superior Court overruled the Circuit Court and accepted Trammell's motion to dismiss. Years of legal wrangling resulted in Trammell prevailing yet again.

Trammell escaped payment of the note. He beat the legal system by having one of the best attorneys in the territory, but it took eight years and five courts to accomplish. Such was the state of territorial justice and the legal tenacity of Nicholas Trammell, whose persistence and endurance in protracted legal matters was consistently on display over his lifetime. His approach to legal matters was similar to the way he treated disagreements over horse racing and gambling—as a challenge between men. No matter what his distractions in the court, during these years Nick Trammell was focused on opportunity. As Trammell already understood, opportunity for him came by way of the roads.

1827—Milam, McKinney, and Parmer

While Trammell secured his new location in southwestern Arkansas for an expanded trading enterprise, other men making bigger plans would be more directly involved with future events in Texas. A series of transactions before the Circuit Court of Hempstead County, Arkansas, demonstrated the early plans of three people later involved in the Texas Revolution. Two of them became signers of the Texas Declaration of Independence. The third led and died in the Siege of Bexar, a key battle for Texas Independence. At a single session of the Hempstead County Circuit Court in May 1827, each of them was involved in actions that gave them oversight for key entry points into Texas. They understood that the road to Texas would pass by their doorsteps and ferry landings, and that road would be very busy indeed. Given their later involvement together in the Texas Revolution, it is entirely possible that the group acted in concert rather than on individual, short-term commercial interests. Seven years before the Texas Revolution, Ben Milam, Collin McKinney, and Martin Parmer were making plans.

Benjamin Rush Milam worked as a land agent for Arthur Wavell, an Englishman granted empresario status by Mexico. The government authorized Wavell to bring five hundred families into northeastern Texas along the Red River on land that was also claimed by the United States as Miller County, Arkansas.[17] Roadways into the prospective colony were critical to its development. Ben Milam appeared before Judge Samuel S. Hall at the Hempstead County Circuit Court in Washington, Arkansas, "praying for the opening of a road from the Town of Washington to Fulton"—the last leg of the Southwest Trail leading to the Red River and Trammel's Trace.[18] The road connecting Washington to Fulton had been improved only two years earlier, but conditions there, leading downhill toward the river, were continually degrading due to erosion and overuse in the spring of 1825. Milam wanted to ensure immigrants to Wavell's colony could make the trip in their wagons to the edge of the United States and to Milam's ferry crossing of the Red River at Fulton.

Collin McKinney was appointed as one of the commissioners to "view and mark the nearest and best way" for the road Milam requested. McKinney and Milam became friends in 1826 when McKinney sought authorization to settle in Wavell's colony.[19] After the two men completed improvements to the fourteen-mile-long road from Washington to Fulton in September 1827, they reported the road ran "to the Forks of Bodarc [*Bois d'arc*] Creek, Thence to the Old Choctaw line, [and] to the cotton Gin on Red River."[20] That road led directly to Milam's Red River crossing at Trammel's Trace. From there, travelers could continue to Nacogdoches or head west on the ridge road on the south floodplain of the Red River toward Wavell's colony and the settlements at Pecan Point. Another route to Texas from Washington, Arkansas, was the old military road to Natchitoches along the left bank of the Red River, down the length of Lafayette County. Men focused on their own stake in the future of Texas were working on the roads and ferry crossings into Mexican Texas. Perhaps they knew, or simply hoped, there would soon be increased traffic. By 1830, Ben Milam was living near Lake Comfort, just a few miles down Trammel's Trace from his ferry crossing at Fulton.

The third character in the Texas drama to come had an even closer connection to Nicholas Trammell. Martin Parmer was in southwestern Arkansas living in the same area as Trammell and his family. After his failure in the Fredonian uprising in Nacogdoches, Parmer settled just beyond the reach of Mexican authority. The nature of the relationship between Parmer and Trammell was curious but apparently allowed them to live in close proximity even though Parmer had killed Mote Askey, Trammell's half brother. Their proximity to each other makes the circumstances of that incident even more mysterious.

At the same session of the court when Ben Milam was licensed to operate a ferry at Fulton and mark the road from Washington to the ferry site, Martin Parmer was appointed as overseer of a thirteen-mile section of a road from Washington to a point between Caney Creek and Parmer's house.[21] Isom Parmer, Martin Parmer's son, had a land grant eighteen miles east of Washington, near a key road crossing. Just as Trammell used his son Nathaniel's name to make the land purchase at the Trinity River, it may be that Martin Parmer purchased land in the name of his son as a solution to his legal difficulties back in Texas.

The March 1828 session of the county court also acted on two other items related to the Parmers that were interesting and ironic. The court approved a tavern license for Martin Parmer, likely kept at the Parmer place to the east on the Washington to Ecore Fabre (Camden) road. Ecore Fabre had river access and was a key port for trade. It generated trading traffic along the road and provided a steady stream of potential customers for Parmer's tavern.[22] In the same session of the court, Isom Parmer was indicted for selling liquor to Indians.[23] The court ruled on two forms of liquor distribution for the Parmers in the same day, one legal and the other illegal.

When Isom Parmer's case came before the criminal court on July 16, 1828, he did not appear but surrendered later to the court's custody. Since Isom surrendered without benefit of counsel, the court accepted his verbal testimony in place of a written affidavit, imposing a bond of $200 for Isom and $100 for Martin Parmer in an attempt to secure their return to court.[24] Two days after posting bond, Martin Parmer asked to be released as overseer of the road passing his house—perhaps a signal that he was already making other plans beyond the boundaries of the Arkansas Territory.[25] As the judge seemed to expect, the Parmers did not appear in court in November 1828. Perhaps an unclaimed letter to Martin Parmer in Chicot County in October 1828 was a notice from the court to appear.[26]

By May 1829, the sheriff again summoned Isom and Martin Parmer to court, and again they did not appear; the Parmers were already on their way back to Texas.[27] Records from the next few years indicate several relocations for Martin Parmer. The 1829 census showed the Parmers in Clark County, Arkansas, but so did the census at San Augustine, Texas. The authorities in Arkansas were not yet aware of the Parmers' notable reemergence in the Ayish Bayou district near the Sabine River in Texas. The Arkansas court was still looking for them in September 1829 when it sent an arrest warrant (alias capias) to the sheriff of Chicot County.[28] The court finally accepted that Martin and Isom Parmer were beyond their jurisdiction. On April 6, 1830, the prosecutor appeared before the court to say he was "unwilling to prosecute this indictment any further" and dismissed the case.[29] Events in Texas pulled Parmer back to the old Neutral Ground, and he would not resist the call.

Changes in Texas
1828—Terán's Inspection of Texas

By 1827, Mexico realized it had effectively lost control of its borders with the United States; not to a concerted military action but by the unrelenting migration of people seeking land. When Gen. Manuel de Meir y Terán was appointed boundary commissioner for the Mexican government in 1827, his appointment was in many ways an admission of that failure. Terán was sent to Texas for an inspection of the province and was expected to return to Mexico City with recommendations for action. Terán's trip to Nacogdoches from Mexico took eight months. He and his party arrived on June 3, 1828, in the heat of an eastern Texas summer. Much of Terán's support personnel, including his heavy coach and instrument wagon, could not cross the flooded Trinity River west of Nacogdoches and was sent back. The group suffered muddy roads, biting insects, and fevers at the end of its long, slow journey along the El Camino Real.[30] Terán found Nacogdoches to be a "mixture of strange and incoherent elements," including peaceful Indians, colonists with some degree of civilization, and more than sufficient numbers of "fugitive criminals, honorable farmers, vagabonds and ne'er-do-wells."[31]

After he arrived in Nacogdoches, Terán sent Peter Ellis Bean to gather information on the settlements at the Red River. Bean traveled to Pecan Point up Trammel's Trace from Nacogdoches with twelve men. His two assignments were to determine the allegiances of the more hostile Indian tribes and to take action "preventing the entrance of adventurers," meaning smugglers and filibusters. After Bean's reconnaissance, Terán learned that there were significant problems with the settlements at the Red River over which he could exercise little control. Mexican authority was much too distant and could do little to address concerns at the northern end of Trammell's road. Terán believed Mexico's bigger problem areas were farther south near Nacogdoches and along the eastern border with the United States.

At the southern end of Trammel's Trace, the population was changing quickly. Terán estimated that there were thirty-four tribes of Indians in the whole Mexican province comprising almost twenty-five thousand people.[32] The population of Nacogdoches in 1828 was around seven hundred, including all of the troops. Only one hundred of that number were women. A count the following year in 1829 tallied 666 persons, of which 99 were slaves.[33] Nacogdoches itself consisted of "forty buildings where English was spoken, the rest are more or less dirty huts with no other floor than the ground."[34] In spite of its deficiencies, Terán recognized the small Mexican settlement of Nacogdoches as the focal point of a "development of the greatest importance to the history of the human species."[35]

1830—Colonization Stopped, Fort Terán

The buildup of settlements near the tenuous border with the United States and the continuing Americanization of Mexican Texas led to major changes in the policies of the Mexican government. The government withdrew authority for immigration and colonization in an effort to stop the overwhelming flow of US immigrants. On April 6, 1830, Mexico reversed its liberal policies and revoked most of the existing colonization contracts, effectively stopping immigration. Only Austin, DeWitt, and de Leon were allowed to keep their colonies. Mexico realized its mistake—all other colonization ended with the enactment of the new law.

The movement of smugglers and immigrants into Texas continued in spite of the reversal of the Colonization Law. When Mexican authorities sought to prevent the passage of immigrants from the United States by adding troops at Nacogdoches, other trails from the Sabine River simply bypassed that village. A large group of families from Tennessee stalled about three miles east of Nacogdoches in 1830 upon learning that they could not pass through without the required passports. Rather than confront the legal issues, they simply made a road around Nacogdoches to avoid the legal entanglements. The road became known as the "Tennesseans' Road" and was used by many subsequent immigrants who did not have the authority to enter Mexico. It may have, in fact, incorporated part of the route between the Trinity River and Trammel's Trace that Nicholas Trammell also used when he left Texas in 1826.[36]

Mexico established new forts in 1831 to guard entry to Texas by way of the Gulf of Mexico or from across the Sabine River bordering Louisiana. One of the forts, Fort Terán, was at a site on the Neches River south of Nacogdoches where three important paths met and led to a single trail heading south. Peter Ellis Bean was assigned to command the fort and began construction in October 1831. The government could not support the supplies needed to maintain Fort Terán, and the last soldiers were pulled out in 1834.[37] Mexico's control of its borders with the United States continued to erode.

Traveling along the Edge of Change
1831—Milam Past Great Raft

Every form of transportation was important in the years of migration toward Texas. When Fulton was laid out in 1819 by Stephen F. Austin and his business partners, part of their dream was that Fulton might become an important crossroads for commerce and shipping down the Red River. The Red River offered opportunity, but the Great Raft, an almost impassable logjam on the Red River, greatly limited trade along that waterway from Natchitoches and New Orleans. Ben Milam became the first captain to successfully maneuver a steamboat past the Great Raft and upriver to Fulton. Milam purchased a steamboat named *Alps* in Natchez, Mississippi, in May 1831 and renamed her *Enterprise*. The vessel could carry thirty to forty tons of supplies but required only a twenty-four-inch draft. Newspaper accounts described how the daring Col. Benjamin Milam "set out from Natchitoches about the 23d of May [1831] with the avowed intention of bringing her through or sinking her in the attempt."[38]

Milam's steamboat towed two keelboats with supplies for Fort Towson up the Red River to the mouth of the Kiamichi River. Seeing Milam's steamboat pulling into the landing at Fulton was a momentous occasion for the residents there. One observer noted that it was "difficult to imagine the powerful effect this circumstance had on the feelings of the citizens. Men, women and children were elated almost to intoxication."[39]

After his trip upstream through the Great Raft, Milam reconsidered the obstacles in spite of the jubilation. He soon announced that he would keep his boat above the raft to transport supplies for the soldiers at Cantonment Towson and for the Indians near there. Perhaps his decision was based more on his desire to avoid another such difficult trip and less on the opportunities for commercial enterprise. In spite of the celebrations, his vessel could not breach the low water shoals of the Red River west of Fulton, and Milam was forced to cancel his ambitious venture. Easier river access for Fulton would not come for many years, but the development of roads across the region continued.

Roads to Jonesborough and Pecan Point

At the end of the western branch of Trammel's Trace farther west of Fulton, Jonesborough became the seat of local government. Red River settlers had been part of the excessively large Hempstead County, Arkansas, but the county seat was far away in Washington, Arkansas. With access to a more local government, they quickly submitted petitions for improvements to both the road from Jonesborough to Nacogdoches and the road from Jonesborough to Washington, Arkansas. In January 1833, the road from Jonesborough along the Red River toward Arkansas was the focus: "petitioners labour under great inconveniences for want of a road or highway to lead from Jonesboro or the court house to the settlement at Pecan Point, leading in the direction of the County of Lafayette (Arkansas)."[40]

No matter how distant the Red River settlements may have been when they were established twenty years earlier, these petitions were testament to a transition that took place. Backwoods trails were no longer acceptable; citizens expected routes to either be improved or abandoned. Settlers who once were forced to use a narrow trace now wanted and needed frontier roads. In March 1833, a list of grievances by thirty-three petitioners from Jonesborough called attention to conditions of the old road: "whereas the present and only road leading from the said settlement situated in the Sulphur Fork Prairies presents many prominent difficulties and aids greatly to the disadvantage of said citizens and travelers in passing to and from the seat of Justice [Jonesboro], difficulties in consequence of their being many creeks to cross which are entirely void of bridges and subject to frequent inundation besides it being a circuitous route adds greatly to the distance, we ask leave to represent from our knowledge of the County that a road can be viewed and marked out from one of the aforesaid named places to the other that would greatly shorten the distance, improve the route and avoid crossing of four creeks in which travelers are liable to encounter many difficulties."[41]

The petition's list of complaints about the present road to the south was reflective of a change in expectations and the increasing number of people using the roads. The "prominent difficulties" the road presented were difficulties only for those who wished to move wagons. Problems faced by small groups of hunters on foot or on horseback were minor by comparison. Crossing swollen creeks without bridges was simply what wilderness travel required. The routes they wished to improve had existed for hundreds of years, with their usual natural obstacles. Animals and people who created the old trails had no civilized destination. Trails turned where the geography changed, and their routes ended at natural places rather than at places where people and property were found. Creeks were sources of water, not obstacles to be avoided. For men focused on land, trade, and settlements, a different route was needed. "The route that we would recommend would commence at Jonesboro, running south or as near so as the face of the country will recommend to the edge of the Sulphur Fork Prairies. . . ."[42]

The court ordered in July 1833 that a road from Jonesborough to Pecan Point be viewed and marked.[43] A road had existed there for many years, but with the ever-changing course of the river, its condition was always uncertain. By September 1833, James Ward, Jesse Moren, John Robbins, George W. Wright, and Joseph Porter were appointed to the task.

A separate request for an improved route south to the Sulphur Fork was approved in 1834 and became a route later known as Dayton's Road.[44] Part of the "circuitous route" toward the Sulphur River that the residents wished to abandon was along the trail called the Spanish Trace. The Spanish Trace (sometimes called the Mexican Trace) was the road created by Spanish soldiers in 1806 when they made their way from Nacogdoches to the Red River to intercept the Freeman-Custis expedition at the Spanish Bluff. It was later incorporated as part of Trammel's Trace. Residents around Jonesborough referenced that road in a request to split themselves off as a separate township of the county. In an 1834 petition to separate a new township, they identified part of the proposed boundary as beginning "about four miles below the mouth of Upper Pine Creek running south with the Spanish trace to the Sulphur fork."[45] The face of the Red River settlements was changing quickly, and so were the roads that Nicholas Trammell and others used to carry on trade.

The Road from Arkansas to Texas

Traffic is hard on any road, and the flood of immigrants using the roads to Texas left their marks. Despite improvements to the Southwest Trail across Arkansas by the military, in 1829 Governor Pope of Arkansas declared the road across his state a disgrace. The governor realized the territory required good roads to support the increasing numbers of people who moved into, and through, Arkansas in search of land and prosperity. Using political tactics not unlike those seen today, the governor shifted the focus from local needs to national ones when he defined the road's importance in military terms. In 1831, Congress appropriated an additional $15,000 for improvement of the Southwest Trail, referring to it as the National Road. Governor Pope was not the only politician to lobby Congress for funds using this naming approach. One journalist commented that "every road thus improved was so called, [therefore] it had little distinction, so the scouts called it the Trail to Texas."[46] During the next four years, Congress appropriated $45,000 for improvements to the Southwest Trail. Additions included a branch of the trail from Washington to the Red River settlements near Fort Towson. Over time, the total amount appropriated grew to almost a quarter of a million dollars.[47] When Congress funded the improvements, the act also laid out specifications for the road: "widening of the road to 16 feet, with all projecting rocks and large stones removed, and for bridges to be built over the larger streams. All brush and saplings six inches in diameter were to be cut to the ground; all trees six to twelve inches in diameter within four inches of the ground; and all larger trees cut within eight inches of the ground."[48]

Despite expenditures, military manpower, and improvements, the Southwest Trail was still a hazardous road where travelers had to deal with "robbers who wait for the unwary."[49] An account of the conditions on the improved road between Memphis and Little Rock puts in perspective the likely conditions on sections of the Southwest Trail that did not receive such resources. The Mississippi Swamp in eastern Arkansas was a formidable obstacle for travel of any kind, even into 1832, when the following account described the difficulties: "We then took the old military road, leading from Little Rock to Memphis, Tennessee. This road lay through swamps, and was covered with mud and water most of the way, for one hundred and seventy miles. We walked forty miles a day through mud and water knee deep. On the 24th of March, after traveling some ten miles through mud, I was taken lame with a sharp pain in my knee. I sat down on a log."[50]

Surveyors laying out the route across Arkansas faced many challenges beyond the difficult terrain. Specifications set by Congress called for a road in this section at least twenty-four feet wide, with timber and brush removed. The road had ditches four feet wide and three feet deep on either side to provide drainage. Low water crossings were improved by puncheons—split timber—laid as a causeway to ease the crossing of wagons. Often called corduroy roads, roads like this were built of 9-foot planks, 2.5 inches thick, placed about 1.5 feet apart. Routes near elevations wound around the hills and dug into the sides of the slope. Oxen pulling scrapers helped make them more suitable for carriages or loaded wagons. In some sections, the government contracted work to residents along the route. One person was assigned work on the road from the six-mile tree to the nine-mile tree.[51]

Any improvement in the roads made the territory more accessible and increased settlement. One of the earliest indications that settlement in the Arkansas Territory was increasing was the use of the early roads as postal routes. Postal roads were hazardous, and the lives of those who carried the mail were in peril due to weather, terrain, and conditions on the trail. Sometimes the postal route was by water. Horace P. Hyde, carrying mail down the White River to Arkansas Post by canoe, was never found and presumed drowned. His canoe was found the day after he left, minus Mr. Hyde but still carrying the mail and about $500 in cash.[52]

The first post road established by Congress in the Missouri Territory on April 30, 1816 ran from St. Louis through Lawrence County, Arkansas, Trammell's original settlement, to Arkansas Post. Post offices across southwestern Arkansas Territory by 1820 were in familiar places operated by postmasters with familiar names. Jacob Barkman at the Clark County courthouse was a member of the large family that migrated with the Trammells. Jacob Buzzard, postmaster at Lost Prairie, in October 1828 sent mail to Long Prairie from his newly established post office. The mail left Lost Prairie every other Friday at 6:00 am, and it arrived in Hempstead County by 6:00 pm the same day. By 1829, post roads were extended to provide weekly mail service between Little Rock and the Miller County courthouse, a 215-mile route by way of Clark and Hempstead County courthouse. Miller Courthouse referred to Jonesborough, the seat of government for the old Miller County, Arkansas. Every two weeks, mail ran from Hempstead County to Natchitoches through Lost Prairie, a distance of 175 miles.

In 1831, the worst conditions on the Southwest Trail were between the Little Missouri River and Washington at the southwestern corner of Arkansas. Within a three-mile stretch containing nine poorly constructed bridges, five were useless due to their poor construction. With the first high water, the planks simply floated away. Subsequent travelers continued the deconstruction of the makeshift bridges. They often removed the remaining boards and placed them back across the streambed without any support or foundation, only to wash away more certainly in the next flood.

George William Featherstonhaugh was a British geographer who was appointed geologist for the US government. He left a record of his journeys across Arkansas in 1834 and documented conditions along the road. As he crossed the Mississippi at Hix's (Hicks) Ferry, his first impressions of the road were good. The road was cut and trees marked to indicate the proper path. Later reports were not as favorable. Farther down the road toward Little Rock, an improvement of settlers who lived off the main road confused and confounded unwitting travelers. Often an alternate route was cut and marked similarly to the main trail but led only to an individual cabin. He also learned that responsibility for the road was left to no one in particular. If a tree fell across the road, rather than move it, travelers would simply go around it—an act that resulted in additional ruts and turnouts. Featherstonhaugh calculated that these turnouts added five miles to his journey across Missouri and Arkansas. The road was still full of rocks, stumps, and mud holes by the time he passed through in 1834. Two times along the road the shaft and front wheels of his wagon were pulled completely off when he tried to free the wagon. Interestingly when he passed through Washington, Arkansas, he found it full of land speculators and others interested in a rebellion in Texas.

Congress authorized additional funds to continue improvement of the trail all the way to Washington, Arkansas. Although the quartermaster corps did not have "an officer of suitable qualifications" available, Lt. R. D. C. Collins was appointed to the project in May 1831.[53] By February 1834, Lieutenant Collins was also assigned to complete the road from Washington to Fulton. In his letter to Governor Sevier, Collins explained that he would not have money to complete this project and needed another $1,000. Sevier passed on the letter to the US secretary of war, Lewis Cass, pointing out that this shortage was due to inaction by Congress.[54] A year later, in February 1835, Congress appropriated more money, and the Military Road finally reached a measure of completion across the plateau of Arkansas.[55]

Improvements to roads and ferries across the Arkansas Territory advanced as the pace of funding and manpower allowed. Since it was US territory, Arkansas benefited from federal funds and the designation of military roads. The US government did all it could with limited resources to support immigration into the West by building roads that supported wagon traffic. Improvements in the road delivering Americans to the doorstep of Mexican Texas and Trammel's Trace reached a new level of accessibility.

More People, More Ferries

The movement of people through southwestern Arkansas on their way to Texas opened the door for the development of even more ferries across waterways on Trammel's Trace and along the Southwest Trail. One ferry operator on the Southwest Trail constructed an improved landing and built an offshoot of the main road to his crossing. In 1832, William Ellis advertised the advantages of his rather ingenious operation: "This road avoids the old site of the United States Road for the distance of two miles and is on dryer and firmer ground than the old road. The subscriber has cut the road at his own expense, and it is now confirmed by the Court. He has graduated the banks of the creek so that wagons, with heavy loads, can pass with ease. The banks are causewayed to the water's edge, so as to be always firm and safe. His boat, recently built, is 60 feet in length and 11 ft. clear in breadth - built of the best materials; and his landings are so constructed, with firm abutments, that the boat serves as a firm bridge, from one abutment to the other in low water, and in high water a safe ferry boat."[56]

To call these improvements to the roads and ferries by the modern term of infrastructure would overstate the nature of the planning that went into the effort. Although the military's improvements to the Southwest Trail were concerted and well executed for the time, it was the push of immigrants toward Texas that created both the political wherewithal for improvements and the commercial opportunism that led to so many ferry crossings. The Fulton ferry operator in 1835 advertised not only his ferry but also the road to the Sulphur Prairies south of the Red River down Trammel's Trace. "We are requested to state, for the information of emigrants to Texas, (and we do so with pleasure), that there is now a Ferry established, and a road opened, leading from Fulton, on the Red river, to the Sulphur Prairies and Texas. This, we are assured, is the nearest and best route for emigrants and travelers going to Natchitoches or that section of the Texas country."[57]

That same year, the sheriff of Lafayette County issued a license for Dooley's Ferry, well downstream from Fulton and closer to Lost Prairie. The road to Dooley's Ferry from Nicholas Trammell's homestead traveled due west, crossing Trammel's Trace near what would become the state line. From Dooley's crossing, travelers could connect to a long-established road along the ridge south of the Red River toward Pecan Point and Jonesborough.[58]

Immigrants who traveled the length of Trammel's Trace from Fulton to Nacogdoches had to cross the Red River, the Sulphur River, and the Sabine River, as well as countless creeks and bayous. Mark Epperson, a Tennessean, received a Texas land grant of a league and a labor at the Sulphur River crossing of Trammel's Trace and started operation of Epperson's Ferry in 1834.[59] Until then, there was only one ferry on Trammel's Trace to make those crossings easier, the one on the Red River at Fulton. Travelers had used a natural crossing near the shoals at the mouth of Anderson's Creek not far from Epperson's ferry as a natural ford for centuries, as well as a similar crossing on the Sabine. Moscoso crossed the Sulphur River twice here in 1542, as did La Salle's surviving crew in 1687.[60]

The network of roads and ferries across Arkansas and into Texas, vastly improved over the time of their earliest use by Nicholas Trammell, was able to support large numbers of immigrants moving south and west. Roads promoted change, and change was coming faster and faster in Texas.

A Business Plan for Texas
1833—Move to Hempstead County

Nicholas Trammell was acutely aware of the conflict that was building across the border in Texas. He had seen the earliest indicators firsthand when he was chased out of Texas in 1826. Although Trammell knew about the rumors of the United States taking Texas from Mexico by force, he saw his opportunities farther from the boundary with Mexico and safely within US territory. A new law regarding public land claims in the United States reinforced his decision and set the stage for a new business plan for the entire Trammell family.

On June 19, 1834, the US Congress enacted legislation that opened the door for an explosion of land claims in Arkansas, many of them fraudulent. The Pre-Emption Act of 1834 granted land claims to anyone who was occupying and cultivating land in 1833. A similar act in 1830 was an attempt to encourage movement to the West from the eastern United States. A failing of the act, or more likely a purposeful obstacle, was that most of those for whom the act was intended could not afford the $1.25 per acre required. As a result, the act simply led to rampant speculation and the use of "floating claims" that could be used to secure land virtually anywhere. Cultivation was required and could include virtually any crop or vegetable. The habitation of the claim required the testimony of disinterested witnesses, but many who proved their claims brought sons and daughters as young as ten years old to testify on their behalf. Some lawmakers believed the law included fraud as part of its design: "It is believed by many that the law itself was intended for fraud, because if it had been, the object of him who drew it, to give the Settler his improvement only, it was wholly unnecessary to give him floating rights, contrary to the very object of all such laws, and if the object of Congress was to prevent any land from being sold for more than a dollar & a quarter an acre, then it would only be necessary to let the lands be entered without offering them at public sale."[61]

Taking advantage of the Pre-Emption Act of 1834, on June 18, 1836, Nicholas Trammell acquired 160 acres in Hempstead County for $1.25 an acre.[62] His claim indicated that he had lived there since 1833, the year after he last appeared on the tax rolls in neighboring Lafayette County. He relocated his family to this land only twenty miles east of Washington, Arkansas, not far from an ancient burial mound. Trammell placed himself yet again at a key road crossing—the same crossing where Martin Parmer previously had his tavern.[63] The east-west road from Washington to Ecore Fabre (which became known as Camden in 1844), a key entry point for trade goods moving west across southern Arkansas, and another road leading from Louisiana to Hot Springs, Arkansas, also passed by Trammell's doorstep. An 1816 map of the region indicates that the road between Hot Springs and Natchitoches was one of the earliest roads in Arkansas.

What Trammell accomplished with this relocation was really quite expansive in a business sense and appears to have been part of his transition to a broader business plan for Trammell and his sons. Mustangs were disappearing, and a market glut sent prices far below a point where profitability offset the danger. Road networks had quickly expanded, and trade goods were more generally available. The tenuous relationships with his Indian partners had likely dissolved. Trammell was fifty-three years old when he moved to Hempstead County, Arkansas. He and his family needed a safer and less physically demanding way of making money than smuggling horses.

Although the road was busy with traffic, there were only four settlers along the Washington-Ecore Fabre road in those years. The well-known and wealthy Magness family from the White River resettled along the road twenty miles west from the Ouachita River. Another twenty miles west was Nick Trammell's place.[64] The Trammells were not after local business; it was the business of increasing numbers of people passing through on their way to Texas. Liquor, horse racing, and gambling were the new trade of the Trammells. Nicholas Trammell and his sons positioned themselves all along the route from the Mississippi River on the eastern boundary of Arkansas to the Red River settlements in northern Texas to take advantage of the roads. The primary road to Texas across southern Arkansas could not be travelled without passing a Trammell tavern or gambling house.

Nicholas Trammell established his base of operations at his new home in Hempstead County, east of Washington, Arkansas. His location was a major intersection of its time with plenty of traffic resulting from trade, migration, and interests in gaming. Trammell's oldest son, Nathaniel, was established on the same road to the east, closer to the trader's port of Ecore Fabre. Nat had been in Chicot County in far southeastern Arkansas at the Mississippi River off and on since 1828. In 1832, another son, Phillip, took up the land hidden in the canebrake that his father formerly occupied in Lost Prairie on the western border of Arkansas, where he developed a successful plantation. Phillip and another brother, Robert, were each fined for operating roulette wheels in Lafayette County, Arkansas in 1833. Their brother, Henry, was similarly located along the road near Dooley's Ferry over the Red River. Robert later relocated even farther to the west along the Ridge Road near the Red River settlement of Pecan Point. Trammell and his sons were stretched across all of southern Arkansas and into northeastern Texas.

Nicholas Trammell's place was so well known among the locals that when a reward was offered for a man who stole slaves, Trammell's tavern was used as the reference point in reporting the crime to the authorities.[65] Hot Springs, Arkansas, to the north of Trammell's new homestead, attracted wealthy patrons from Louisiana who believed in the healing benefits of the hot mineral water. Farther north along the same road, Trammell's uncle, David Trammell, kept a tavern called Fair Play at the place where the road to Hot Springs joined the Southwest Trail between St. Louis and Fulton.[66] The route used to get to Hot Springs from Natchitoches passed right by Nicholas Trammell's crossroad home. "Nick Trammel, at his residence on the Terre Rouge, also entertained the visitors and health seekers for the South on their way to Hot Springs. These planters from the South had plenty of money, and 'Nick' at his place and 'Polly' at her place got much of their specie [money]."[67]

Polly Vaughn may have been as much of a character as the old smuggler Trammell. An 1823 version of a pre-nuptial arrangement guaranteed her payment of twelve dollars for each month she lived with Stephen Vaughn and possession of his assets in the event of his death. A house and lot in Washington, three horses, furniture, and other material goods were on the list.[68] Her services and establishment were so well known that Stephen F. Austin trusted Polly Vaughn to care for a trunk full of belongings he wished kept secret. The contents of the trunk are unknown.[69]

Gambling as a Family Business

Trammell gaming interests included roulette, racing horses, and other ventures. Gambling operators who operated roulette wheels had the odds more in their favor than dealers of the more popular and pervasive frontier game called faro, which began to spread around 1825. Faro was a banking game, somewhat like baccarat, where play was against the dealer. Faro dealers often traveled from place to place with portable gaming equipment easily set up at taverns and quickly dismantled for a hasty exit. Nicholas Trammell and his sons' gambling interests may have included faro tables in Arkansas.[70]

Travelers losing money at Trammell's tavern spread their stories of gambling losses. Years after his death, Trammell's reputation was embellished with rumors that "the footsteps of several lone travelers had been traced as far as Nick Trammell's cabin where they had vanished forever."[71] Killing off customers would not have been a good business plan, and it is likely that the tall tales were just that.

Gaming of all kinds was ingrained in the frontier mentality at the western edge of the United States. During the presidency of Andrew Jackson, which began in 1829, a view of gambling as sinful and unhealthy began to conflict with the realities of gaming in Trammell's part of the country: "A gambler, after reaching his highest state of excellence, is generally apt to retrograde rapidly. From faro he will visit the race-course and cock-fightings, where he will get with men who are reckless and vicious as men can be, and still be at large in the community."[72]

Citizens complained that "every reasonable and rational amusement appeared . . . to be swallowed up in dram-drinking, jockeying, and gambling," all activities in which Trammell and his sons engaged over many years.[73] On November 10, 1829, the outcry against rampant gambling in the Arkansas Territory resulted in the passage of "an act to prevent the evil practice of

Gaming." The act outlawed wagering at cards, dice, billiards, faro, cockfighting, and various other gambling endeavors. Where the Trammells lived at the edge of the nation, different rules applied and the impact of the law seemed not to penetrate. With Trammell family members operating taverns and gambling operations in Lost Prairie, along the Washington to Ecore Fabre Road, and on the Southwest Trail at Hot Springs, the family had a significant presence along the early routes. The chosen profession of the Trammells once again put them on the side of those of questionable character and scruples. They engaged in business activity that was illegal but carried on just beyond the reach of the law.

Even the presence of the US military could not deter the gambling spirit. In fact, it often supported those diversions. Fort Jesup at Natchitoches was meant to provide some semblance of law and order in the borderlands along the Sabine River. Instead, it attracted goings-on that could hardly be considered lawful or militarily beneficial. Two miles west of Fort Jesup on the El Camino Real nearer to Nacogdoches, a "Shawneetown" sprung up to supply the troops. It was not unusual for saloons, gambling houses, racetracks, and other leisure activities to lure the attention of soldiers away from the loneliness of their posts.

Nacogdoches was also a haven for gamblers more than eager to welcome a newcomer to Texas. Although the local council in Nacogdoches tried to control gambling through fines on gamblers and operators, their efforts were largely unsuccessful. The unwritten rules often left the unwitting relieved of their stake. One gambler, an unnamed future signer of the Texas Declaration of Independence, spotted an easy mark that appeared to have money and directed him into a game. After the newcomer sat down, the future patriot went out to look for others to fleece: "When he returned the game was over and the clique dividing the spoils. The steerer demanded his share. 'Why you was not in the game,' they contended. 'The hell I was not; didn't I find him first?' and backing his claim with a pistol he secured his share."[74]

Societal pressures reflected in the antigambling legislation drew tighter around Trammell in 1835 with the formation of an Anti-Gaming Society in Hempstead County, Arkansas.[75] Editorials supporting such movements focused on the impact of the vices of gambling, seen by Arkansans as contributing "more to degrade us in the eyes of the world than any other."[76] Even gamblers themselves participated in the outcry, hoping to banish the "dishonest" gamblers from a sport that was often little more than another form of theft. The antigambling societies hoped to instill fear in those who gambled, as well as those who sponsored the operations. Their pledge was to use their influence to displace elected officials who supported or participated in gambling and not patronize any "tavern or public house, or the keeper of any tavern or public house, who shall keep or entertain any professional gambler in his house."[77] If the Hempstead County Society acted fervently on that pledge, it was apparently of little concern to Nicholas Trammell and his sons.

Improvements to the roads passing by Trammell's place may have had much to do with the success of his business. Commissioners were appointed in January 1834 to mark out and improve a road from Trammell's to the Little Missouri River and on to Greenville in Clark County in the direction of Polly Vaughn's tavern. By December 1834, the US House of Representatives heard Congressman Sevier's appeal for $20,000 to improve roads along which the Trammells focused their farming, taverns, and gambling operations. Sevier's request focused on an east-west road that stretched from Jonesborough on the Red River to Chicot County in southeastern Arkansas, which crossed right in front of Trammell's tavern. He also requested improvements to the national road from Little Rock to Washington, the Southwest Trail that led to the Choctaw line.[78] Trammell's reputation and the significance of the road past his door was broadcast on a national level when the Arkansas Territorial Assembly petitioned the US Congress on October 24, 1835, for funds to improve the Washington to Ecore Fabre road. "Your memorialists the General Assembly of the Territory of Arkansas respectfully represent that the sparce population of our Territory renders them unable to open roads for their communication with the neighbouring States. That large bodies of the publick lands of the best quality are unavailable to the Government by the difficulties presented to the emigrant in reaching them with his family and property. That no section of Country illustrates the necessity of opening roads more than that which lies between Columbia on the Mississippi river and the Town of Washington in the county of Hempstead crossing Bayou Bartholomew at or near John Stuarts and the Washita at or near Ecora Fabra and intersecting the Military road leading from Natchitoches La. to Fort Towson near Tramels a well known stand on said road."[79]

All roads led to Texas, and one of the primary roads across Arkansas went past the "well known stand" of the Trammell family. Increasing migration led to more customers and more people, wagons, and livestock moving to the south and west. While there were favorable business conditions for Trammell and his sons, the country's attention was focused even farther south and west in the territory of Mexican Texas. Nicholas Trammell lived near the edge of an unfolding struggle.

Trouble Brewing in Texas

Settlers in Pecan Point and Jonesborough were often forgotten or ignored by the two governments in dispute over the boundary. The residents there wanted to become Mexican citizens but continued to behave as if they were still in the United States. They served on Arkansas juries and borrowed from banks in the United States, but when tax time came they claimed Mexican citizenship.[80] Benefitting from the lack of agreement and their distance from any ruling influence, they liked their ability to change allegiances to suit their needs but did not like the growing sense that the United States was abandoning their interests. The affirmation of the treaty rights of the Choctaw to land the Anglos occupied was the pinnacle of their disgust.[81]

Soon after the Fredonian Rebellion in Nacogdoches, leaders of the Pecan Point group expressed their common desire in a letter to Jose Antonio Saucedo, governor of the Mexican Province of Texas, to submit to Mexican governance. Saucedo's proclamation of June 12, 1827, in response to that request, made clear that a more liberal Mexican government would willingly accept the Anglos into Mexican Texas with conditions. The question of whether the Red River settlers were Americans settling illegally on Mexican soil or were Mexican residents of the Province of Texas could not be clearly answered. Even though the Adams-Onis treaty laid out a boundary between the United States and Spain along the Red River, by 1828 it still had not been marked or surveyed.

Neither group of settlers and immigrants was happy. In a gesture of the dissatisfaction with their governance by Miller County, Arkansas, the citizens of Jonesborough burned the courthouse on November 5, 1828, destroying all the records.[82]

1834—Almonte's Inspection

The unrest among the Red River settlers was a microcosm of the confusion over rights, boundaries, and ownership that prevailed in Texas. Mexico's hold on Texas was slipping. Threats to Mexican sovereignty in Texas continued to the point where even the most resolute officials understood Texans were prepared for revolt. In 1834, Col. Juan Nepomuceno Almonte was sent to Texas for another inspection of the tensions along the border, similar to the one by Teran six years earlier. Almonte's assessment did not reassure the Mexican government.

Almonte found the state of Mexico's practical abandonment of Texas appalling. The vast expanse of territory was secured by only one hundred soldiers in all of Texas.[83] Mexico had virtually given up its claim on Texas by default. Around Nacogdoches, peaceful Indians kept the more hostile tribes at bay, but in central Texas, the warlike Comanche reigned.[84] Travelers along the El Camino Real through Nacogdoches were in jeopardy "either from attack by the savages who are currently committing outrages or from assault on the part of the settlers, whose conduct we cannot trust."[85] Trade along those roads continued unabated. In a single year, eighty thousand deerskins were shipped from eastern Texas as a result of trade with Indians from the region.[86]

Almonte noted the only military presence for the United States along its border with Mexico was at Fort Jesup west of Natchitoches and Fort Towson north of Jonesborough. Fort Jesup was on the busy El Camino Real. Three hundred infantry troops and six pieces of light artillery stationed there were not behind any fortification. Fort Towson had a similar contingent of troops guarding the northern border along the Red River. Almonte believed Mexico should match those fortifications with Mexican detachments at the Red River and the Sabine River. A threat of military action by US soldiers massed outside the borders did not create problems for Mexico. The tinderbox for revolution was already being prepared from within its borders. All it needed was a spark. On December 10, 1832, that spark ignited Texas for the first time.

1832-1836—Houston, Crockett, and Bowie Enter

In the winter of 1832, the former governor of Tennessee, the adopted son of a Cherokee chief, and an incorrigible drunk entered Texas—all in the singular persona of Sam Houston. Like many others from Tennessee, Houston did not enter Texas via the Red River at Fulton but farther upstream at the Red River settlements full of other Tennesseans. Ben Milam, who cleared the path for Houston's entry in many ways, was there with Houston and summoned John Ragsdale to bring his ferry across to the north side of the Red River, opposite Jonesborough, so Houston could cross.[87] On the south side of the river, Houston found a respectful welcome when he visited the home of Daniel Davis. Houston was not sure what would become of Texas, but he had been certain for some time that he wanted to be a part of the outcome. Traveling south down Trammel's Trace, Houston arrived in Nacogdoches where he established a law office in April 1833. He stayed only a few months, long enough to gather firsthand information about the conditions in Texas, and then left the region for the eastern United States. When Houston returned to Texas in December 1834, he came back down the Southwest Trail and stayed at the Traveler's Inn in Washington, Arkansas. From that outpost, Houston prepared himself to become a key figure in the coming battles for Texas' independence from Mexico.[88]

Houston traveled the Trace again in February 1836, this time on a mission to shore up support among the Indians for the coming revolution. Houston met William Fairfax Grey, George Hockley, and Alexander Horton along the El Camino Real west of the Neches River while on his way to hold a treaty with the Cherokees, Shawnees, and other Indians settled in Texas, north of Nacogdoches. Houston no doubt knew Trammel's Trace well as a result of his many movements between Texas and the United States.

David Crockett followed a similar geographic path to Texas but came for different reasons. Crockett, the "go ahead" man, became used to public attention to his homespun ways but never could seem to make his fortune. When Crockett learned about the land available in Texas, he saw opportunity in the form of a four-thousand-acre headright of land. During an election in Tennessee, Crockett famously told his electorate that if he lost, "they might all go to hell and he would go to Texas," and that is what he did. By November 1835, Crockett's journey south was celebrated in Little Rock, and then he continued through Washington and Fulton on his way to Texas. Rather than heading directly to Pecan Point, he ventured south toward Lost Prairie and crossed the Red River there, likely at McKinney's Landing.[89] By early December, Crockett traveled the road to Pecan Point along the Red River on the south side, pausing long enough for a buffalo hunt in the Sulphur prairies. The familiarity of fellow Tennesseans and the fertile land along the Red River caught Crockett's attention. Had he survived the battle at the Alamo, Crockett would have been a Red River settler himself. Crockett wrote to his wife: "I expect in all probability to settle on the Bordar of Chactaw Rio of Red River that I have no doubt is the richest country in the world good land and plenty of timber and the best springs & mill streams good range clear water—and every appearances of health game plenty. It is in the pass whare the Buffalo passes from North to South and back Twice a year and bees and honey plenty."[90]

On his way to his destiny at the Alamo, Crockett traveled down Trammel's Trace to Nacogdoches, arriving there on January 5, 1836. Cannons were fired for his welcome, followed by a dinner and dance organized in his honor. The excitement of being in a "bran fire" new country buoyed his spirits and spurred his determination to be a part of the coming revolution. Crockett took his oath of allegiance to the fledgling government of Texas in Nacogdoches. It was in Nacogdoches that he also left behind the trail that would have taken him to his hoped-for home along the Red River were it not for fateful events to come. Crockett would not retrace his steps over Trammel's Trace. His and Texas' fortunes would be bound together forever at the Alamo.

James Bowie traveled Trammel's Trace many times. His land frauds across three states, his participation in earlier skirmishes, and his involvement in the slave trade provided him with intimate knowledge of every backwoods trail. Bowie made his way to Nacogdoches in July 1835, where he actively solicited troops to come to Texas. Bowie's reputation put him at the head of a swiftly organized militia, which brazenly stole its arms from the Mexican munitions warehouse in Nacogdoches. Bowie's bravado and drunkenness led him to demand Sam Houston's horse when they were both in Nacogdoches, a demand to which Houston agreed on the advice of those who knew Bowie well.[91] Everyone knew Bowie was a fighter, and he had the scars to show for it.

A rebellion by Anglo Texans against Mexico was increasingly certain. Concerns over the loyalties of the Indians in Texas when war began were paramount. Sam Houston had deep, personal ties to the Cherokees, and, with Houston's blessings, Bowie visited the villages of the tribes in eastern Texas to ask for the help of their warriors, or at least to remain uninvolved when war came. During Bowie's tour of the villages north of Nacogdoches, he found their people dancing and drinking. They were so drunkenly impaired that the impatient Bowie would not wait for them to get sober. Bowie returned to Nacogdoches without an agreement.[92]

Sam Houston, David Crockett, and James Bowie each rode along Trammel's Trace on their way to destiny at the Alamo. Men who were vilified, sued, disgraced, and in debt in the United States became Texas heroes. Events in Texas moved quickly toward revolution. A consultation in November 1835 set the stage for a Declaration of Independence from Mexico. The following month, Sam Houston broadcast for volunteers to join the Army of Texas.

The turmoil in Texas was change on a grand scale, but it also changed the plans of a Trammell relative. One of Nicholas Trammell's cousins from his mother's Maulding side of the family stayed at Trammel's place in December 1835. Rather than continuing on his way to Texas, a territory unsettled by the coming war with Mexico, Presley Maulding stopped to stay for a short time with Nicholas at his place on the Washington to Ecore Fabre road. Maulding's brother-in-law, Martin Poer, wrote to his father in Tennessee, telling him about his journey. Like many others traveling the Southwest Trail, Maulding and Poer had difficulties crossing the Mississippi Bottom on the western side of the great river: "We had a dreadful time in the Mississippi Bottom. It commenced raining the next day after we crost the River which made the Road very wet. We past a great many wagons on the road while we were traveling. We out traveled many People that has been through the Bottom this season. I would advise every Body that moves to Texas to go by water. They have no idea of the danger and exposure of people and horses. For my own part I have been exposed on the road more than I ever was for the length of time in my life. For several days in the water and mud up to my nees and very often deeper from morning till night. My load was too heavy to allow me to ride and drive. . . ."[93]

Maulding lived on Trammell's farm and spent part of his time in the woods hunting for bear and deer. Poer rented about six acres of land four miles from the tavern, putting up a little camp. Poer expressed the hope that he could make improvements on the government land and profit when he sold it. Like many others huddling just beyond the borders of Texas, Poer planned a move to Texas after the revolution ended.

The battle for Texas Independence is well documented and will not be retold here. When the Texians defending the Alamo were overrun and killed on March 6, 1836, eight men from the Red River settlements died. Eight men of the over 150 who died were from the assemblage of "bad men" along the Red River. Two months after the Alamo fell, a panic called the Runaway Scrape set colonists rushing along the El Camino Real and any other road toward the United States with anything they could carry, attempting to escape a rumored Mexican massacre. Disabling weather, sickness, hunger, and death were the realities that met them.

Luckily, Sam Houston's resolve to fight on matched the opportunity. His diverse assemblage of Texians brought the revolution to a successful end at San Jacinto when they captured Santa Anna and his army. Texas was no longer part of Mexico and became an independent Republic. A later retelling of Trammell's participation in the fight for independence had Daniel

Davis and Nicholas Trammell gathered at the Trinity River crossing with fifteen to twenty men from Pecan Point preparing for battle. The story was that Davis and Trammell did not arrive there until the afternoon of the final battle at San Jacinto.[94]

James Winters, one of the soldiers at San Jacinto, traveled to Texas down Trammel's Trace from Tennessee to join Houston's army. After his engagement in a small skirmish at Lynch's Ferry, just north of the final battlefield of San Jacinto, Winters was happy simply to have time for a meal. He contentedly reported that "after the first onslaught the Mexicans fell back and we got our breakfast."[95]

On the battlefield at San Jacinto, James Winters got his breakfast, Sam Houston got a nation, and Nicholas Trammell's road changed character yet again.

8

1836–1844
Another New Nation for Texas

Texas is a whole other county.

—THEODORE PAVIE[1]

After winning its independence in 1836, Texas became a country separate from both Mexico and the United States. And the trade route Nicholas Trammell used to smuggle horses changed its nature once again.

Trammel's Trace had not been a hidden trail for fifteen years, since migration increased after Mexican independence from Spain in 1821. From 1823 to 1830, Anglos trickled into Texas at about a thousand a year. They settled primarily along the borderlands or in the colonies of Central Texas. Even after Mexico stopped colonization in 1830, the numbers grew to three thousand a year. By 1835, the Anglo population of Texas, including their slaves, was estimated at about twenty-five thousand. Anglos outnumbered Mexicans in Texas by ten to one, even though Texas had been a Mexican province.[2]

The number of people in the region transected by Trammel's Trace also grew at a rapid pace, overtaking the country and the roads. Most of those who came were farmers from the south, there for the land. Many of them were poor, fleeing debt or the courts. They were not like what one would expect back in the more settled territories. Even a wizened soldier visiting Nacogdoches was surprised by their rough nature: "The dress of the Americans or the setters from the united states differ from what you are accustomed to see part of thear apparel consists of 1 or 2 braces of pistols or large knife a sword and gun a furious look and frequently use thear implements on each other."³

In spite of that reputation, even the relatively isolated settlement of Nacogdoches was able to call up a sense of the refinement when needed, as the celebration in honor of David Crockett demonstrated. William Fairfax Grey, while on his way through Nacogdoches in February 1836 on behalf of investors seeking cheap land, had time to enjoy the scenery along the route of Trammel's Trace just north of the post: "Rode out in the evening with some ladies a few miles up the Pecan Point road, north; a beautiful carriage road, over sand and red land, leading up the Banito [Banita Creek]."⁴

This more improved part of Trammel's Trace was within just a few miles of Plaza Principal along the Calle del Norte in Nacogdoches. Apparently, a war with Mexico not too many miles to the west was insufficient cause to interrupt a carriage ride with the ladies.

After the birth of the Republic of Texas in 1836, the primary focus of virtually every resident and many interested parties still in the United States was land. On December 22, 1836, the Texas Congress passed a law establishing a General Land Office for the Republic. Land offices finally opened across the Republic of Texas in February 1838, creating a rush to Texas for land that was for all intents and purposes free. Confusion, conflict, and unscrupulous dealing were inevitable. The statute created eleven land offices within the bounds of the republic. Two of them covered the area of the state traversed by Trammel's Trace. Land Office No. 2 was located at San Augustine for the counties of San Augustine, Jasper, Sabine and Shelby. Land Office No. 1 was located at the house of George Wright on the Red River. That office included all the land within the specified boundary: "the line to begin at the mouth of the Sulphur fork of Red River, and run up that river to the crossing of the Trammel trace, thence on that trace to the Sabine river, thence up that river to its source, thence due north to Red River, thence down that river to the beginning."⁵

When the land offices were established, the Texas Congress left a relatively large area north of Caddo Lake and east of Trammel's Trace without a designated land office. This oversight not only required Red River residents to travel great distances in order to make claims, but a large area along the border in northeastern Texas was not specified in either district. To address this gap, petitioners from the former Miller County area near Louisiana and Arkansas asked that a new county be established to place related administrative functions closer to the signatories and to assign a more conveniently located land office.

Not everyone caught up in the rush to Texas remained in the new Republic. Some of the settlers who left the Red River settlements for Texas decided to retrace their steps up Trammel's Trace and return to Pecan Point. Daniel Davis moved from his land in the canebrakes along Ayish Bayou, near present San Augustine, Texas, back to Jonesboro. The rapid pace of immigration put increasing pressure on game, on fur trade, and on the herds of mustangs that once provided plentiful spoils. On Daniel Davis's 1836 return trip north past Blossom Prairie, about fourteen miles west of Clarksville, there were only fifteen wild mustangs where herds once thrived.[6]

Horse Racing

As the numbers of mustangs available on the prairies shrank ever lower, frontier interests in horses shifted from capturing them to racing them for a purse. Horseracing all over the Arkansas Territory was a well-documented part of the culture of the frontier. Little Rock had a racetrack by 1825, and Long Prairie, where Trammell lived, started a jockey club in 1826. Lafayette County had a track by 1832, and the track at Washington, in Hempstead County, was operating by 1834. Surrounding counties of Clark, Crawford, Independence, and Jefferson were operating tracks by 1836. That same year, Arkansas became the twenty-fifth state in the union and had a population of just over fifty thousand people. Many small towns along Trammel's Trace and the El Camino Real in Texas also had racecourses, including San Augustine, Nacogdoches, Crockett, Clarksville, and Boston.[7]

Racing was not limited to the outer edges of frontier culture. A Methodist missionary in Louisiana was appalled to find a track immediately in front of a church and that the priest himself owned a racehorse.[8] As one advertisement noted, the racecourses and jockey clubs were "for the recreation of those who devote a portion of their time to the scenes of high life."[9] They were also a place to find Nicholas Trammell and his sons.

Only one account in the words of Nicholas Trammell survives history and speaks to us from the record. Fittingly, those words focus on a horse race. It is not surprising that Trammell engaged in a lively challenge from time to time, given his family history in racing and gaming. The account of his son's horserace in Nacogdoches and his attempt to recover the pig he won demonstrated Trammell's willingness to place an impromptu bet and to persistently pursue his winnings. Shortly after Trammell established himself along the road from Washington to Ecore a Fabre, a horseracing challenge between two other parties became public. Trammell's published response to the banter between two other men is a testament to his enjoyment of a challenge.

On June 14, 1836, a racing challenge was published in the *Arkansas Gazette*, a four-page newspaper that contained news of the day along with editorials and letters for public consumption. The challenge did not involve Trammell directly, but he clearly took an interest. On that date, a letter addressed to a Mr. Hawkins appeared, submitted by the rather wordy John Loring. Loring boasted that his horse, Sir William, could beat Hawkins' horse, Uncas.

> A BANTER: I can beat Uncas either at this place [Batesville], or Little Rock for $500 or $1000, 2 miles and repeat upon the following conditions. If you prefer coming to this place, I will pay all necessary expenses therefore, otherwise, I will come to Little Rock, at your expense. I will carry 80 or 150 lbs. I would not make this obstacle, but I am a stranger in Arkansas and am unable, therefore, to obtain riders of every weight, upon whom I can rely. I will run on or before the 15th of October next, at either of the above mentioned places that may be selected and specified by yourself, if done within two months from this date. To make short of a long story, Sir William can beat Uncas upon fair terms for either of the above amounts. Should you select this as the place of contest, after agreeing to the above propositions, and should there be any by-laws of this track, to which you may demur, I will be governed by the laws and regulations of the Little Rock track. I will subjoin for your information that the members of the Batesville Jockey club, will meet on Wednesday next, for the purpose of putting the Turf in order for Horses of the highest blood.[10]

Batesville was the seat of justice for Independence County, Arkansas, on the navigable part of the White River. It grew rapidly and became the largest community in north central Arkansas since the time Trammell had first lived there in the early 1800s. Batesville's jockey club set September 29 as the opening day of its regular fall meet.[11] Trammell was not directly involved in the challenge from Loring, but something about the circumstances caused Trammell's racing instincts to react when he read Loring's banter. Either to defend Hawkins or take on the confident Loring, Trammell published his reply to Loring's challenge in a notice that appeared in the *Arkansas Gazette* on July 19, 1836.

> Hempstead County, July 4th, 1836
>
> To Mr. John Loring,
>
> SIR—From your communication to Mr. Hawkins of Little Rock, you seem to complain of your misfortune in not obtaining a race with him - and make banters for him to accept, which, it seems, you know could not be taken up by him. Now, sir, permit me to say, that I will take your banter upon the terms specified in your communication to said Hawkins; only that I will require you to come to the Hempstead Races, and all the expenses consequent upon the trip, with all other expenses appertaining to the stable and keeping of your horse, shall be paid, and will run you for from one to two thousand dollars, and thank you for your custom.
>
> I am sir, with respect, your, &c. Nicholas Trammel
>
> P.S. Please inform me whether you come or not.[12]

The Hempstead racetrack was located east of Washington, Arkansas, just across Town Creek. The circular half-mile track was used for regular heats of a mile or two.[13] Not only did Trammell come to the aid of the apparently unavailable Mr. Hawkins with his public challenge, but he doubled the bet that Loring laid out in the newspaper. The tone of Trammell's postscript to Loring has the sound of someone exceedingly confident. Trammell essentially asked the confident Mr. Loring to put up or shut up.

People with whom Trammell associated also held a fondness for horses. Daniel Davis, another Pecan Point settler, was a horseman who liked to race for sport and spoil. Andrew Davis described his father's passion for horses: "My father was a great admirer of a fine horse and had a good stock. You could not give him a horse that had the Mexican brand on it. But a few years before his death, he visited the races at New Orleans and bought a mare off of the turf that was getting too old to have speed. He paid $300 for her and sent her through a perfect wilderness from Red River to San Augustine, a distance near 200 miles, and bred her to Earl's Old Packet, a very noted horse in that region, paying $50 for the colt."[14]

Davis made this trip to San Augustine along little more than an old Indian trail. Trammell's trail led him to Nacogdoches, where he turned east on the El Camino Real to San Augustine. It was unlikely that the long, difficult trip across the "perfect wilderness" was made solely to breed a mare, but any other business Mr. Davis transacted is left to our curiosity and imagination.

No records provide insight into Trammell's specific business interests in racing, but his chosen business was part of a significant and lucrative trade in horses that potentially generated substantial investments and rewards. A man named Farrar of St. Francisville, Louisiana, took four racing horses to Texas in the summer of 1839 and sold them for $11,000.[15] Gambling of all types remained a widespread form of entertainment in the frontier, but horseracing had the broadest appeal and the biggest prizes. Published challenges ranged up to $10,000 for a single race. The popularity of racing waned during the 1840s, when antigambling sentiments imposed restrictions, or even outlawed, gambling and horseracing. The Hempstead racetrack where Nicholas Trammell no doubt won and lost thousands of dollars fell into "innocuous desuetude" after 1843.[16]

Roads in Texas

Although the formation of the Texas Republic seemed to solidify a new governance for Texas, the mere existence of the new nation had little impact on the condition of its roads and trails. In Columbia, fifty-five miles northwest of Houston and the seat of Texas government in 1836, a man died when he fell across a stump left in the middle of a road. A newspaper lamented similar conditions in crossing a street in Clarksville, in northeastern Texas near the Red River: "you bark your knees successively against half a dozen [stumps] before you reach your destination; after which you offer up sincere thanks to the giver of good that you did not break your neck, or fall in the mud, both very possible occurrences. With a most providential regard for their successors, the first settlers of the city, left the stumps of all the trees they felled just high enough to strike a man's knees, for which considerate kindness we return thanks with tears in our eyes."[17]

Increased settlement meant increased use of the roads, including Trammel's Trace. By April 1837 mail routes were open and running across the new nation of Texas. The Republic of Texas officially recognized part of Trammel's Trace when it instituted a mail route from Nacogdoches to Jonesborough. The new Texas Congress enacted a law in December 1837 to "establish as soon as practicable a mail route from Nacogdoches, via Epperson's Ferry on the Sulphur Fork, to the county seat of Red River County, Jonesborough." Mail deliveries were made along the route once every two weeks. The location of Mark Epperson's ferry on the Sulphur River made it a good site for a post office. The decision was simplified by the fact that there was no other suitable road available.

Later mail routes followed paths laid out along Trammel's Trace and other connecting roads used by the earliest Red River settlers. Once-a-week mail service along the Red River from Myrtle Springs, Texas to Fulton, Arkansas, was established in January 1839. Myrtle Springs was a small community in Red River County at the time and is now Bowie County. At its eastern end, the mail route connected with Trammel's Trace south of Fulton.

Changes to the Texas Republic's mail routes over time highlighted the swift development of new destinations and roads to reach them. A single mail route from San Augustine to Port Caddo, just west of the boundary with the United States near Jefferson, was rerouted to include new settlements at Shelbyville, Pulaski, Elysian Fields, and Marshall. The Nacogdoches to Epperson's Ferry route was altered to make stops at Marshall and Daingerfield, just off the original path of Trammel's Trace. From Epperson's Ferry to Jonesborough, a new road ran through DeKalb and Clarksville in Red River County. Prominent Nacogdoches resident Adolphus Sterne considered bidding on the Epperson's Ferry mail contract along with a man named Matt Simms, but a Mr. Caldwell from Fannin County got the contract. Newer roads developed as offshoots of Trammel's Trace. Roads to new towns had to be cut through the wilderness of eastern Texas and, once cut, had to be well traveled in order to remain clear.

The pace with which settlements and towns in Texas grew during this period resulted in rapid changes to patterns of travel and communication. Hughes Springs was founded in 1839, Jefferson and Marshall in 1842. New counties were formed from others—Harrison County in 1839, Bowie County in 1840, and Rusk County in 1843. On early maps of the Texas Republic, Trammel's Trace appeared to curve around and avoid the new settlements of Jefferson and Marshall. In fact, those communities did not exist when Trammel's Trace was the only road in the region. Trammel's Trace was the path Indians took in search of game and making their way between villages. Generally, it followed the terrain. Industrious people with axes would soon build new roads directly connecting new Texas settlements. New roads quickly crisscrossed the sandy hills of eastern Texas, making travel more direct. The abandonment of Trammel's Trace as the primary route for travel across northeastern Texas was beginning.

Settlements sprang up and then sometimes disappeared rather quickly. The settlement of Jonesborough had existed on the Red River since 1820 and grew to prominence as a result of its key role in the trading culture. A town plat showed the locations of a tavern and mercantile, and as late as 1837, Jonesborough had forty to fifty houses. Just as quickly, fortunes turned. By 1842 the constant flooding and changing course of the river drove businesses and homes farther south. The route from Jonesboro tracked east-west following the Ridge Road along the floodplain south of the Red River. Beyond the initial floodplain of the river, a second bottom was bounded by bluff higher than the first banks, more easily worn down by the river's frequent overflow. Even the higher road was underwater in places during excessive flooding in 1843. Flooding completely covered the entire Jonesborough Prairie that year. During that year's flooding, only the tops of Indian mounds remained as visible reminders of the landscape.[18] Jonesborough soon disappeared from the map, swallowed by the sediments of the Red River and preempted by newer settlements more favorably located.[19]

Ferries

Swimming a horse or floating a wagon across a river became an insufficient means to manage the transportation needs of the new republic. Ferrymen had known this for years and established themselves at key crossings. Ferry crossings named for their operators were found at every major crossing along Trammel's Trace soon after the end of the Texas Revolution. Emerging county governments quickly established rules for ferry operators and set rates they could charge for crossings. Along the El Camino Real at the Trinity River, where Nicholas Trammell once operated a ferry, there were well-established ferries in place for many years after his exit from Texas. West of Nacogdoches, the approach to the Trinity River was through a boggy, miry prairie several miles across in an area subject to overflow when the river was high. Between steep banks twenty-five feet above the river, Nathaniel Robbins operated his flat boat ferry. Robbins lived on the east bank of the Trinity, perhaps near the same place previously occupied by Nicholas Trammell. A well-known Nacogdoches businessman, John Durst, operated a ferry across the Angelina west of Nacogdoches on the El Camino Real. His fee was one dollar, where most other crossings on that road cost only twenty-five cents.

Business opportunity and geographic possibility resulted in several choices of ferry crossings near Fulton at the Great Bend of the Red River. The Fulton crossing had ferry service since 1821 but had shifted among various locations depending on the changing nature of the river. Pope's Ferry crossed the Red River near a shoal at the mouth of Little River just a few miles upstream. Even farther up the Little River was Allen's Ferry. Ferry operators favored locations in the deeper water just upstream from natural low-water crossings. New crossings were never far from where the buffalo and Indians found their own way for hundreds of years. To the south of Fulton down the east bank of the Red River, Dooley's Ferry became a well-used alternate crossing. Roads toward Dooley's Ferry from Fulton on both sides of the Red, one passing through Lost Prairie, had been established for some time. The road from Dooley's Ferry toward Texas crossed Trammel's Trace just west of the Texas-US border and merged into the Ridge Road toward Jonesborough.

The businessmen running the ferries were dependent on word of mouth and newspaper advertising to direct Texas-bound travelers on their way. The operators of Dooley's Ferry in 1837 learned that "travelers and movers to Texas" were being told that the road to Dooley's Ferry from Fulton was nearly impassable. They were probably right in assuming that information was being spread by competing operators. To combat the rumors, the Dooley's Ferry licensees, William M. Burton and William Cunningham, placed an ad in the *Arkansas Gazette* in September 1837.

> We will further state that the accommodations on the Dooley's Ferry road are as good, if not better, than on either of the other routes—that the road is, in our opinion, and that of many others, decidedly better and nearer—that the crossing of the river is entirely safe and expeditious - there being two good ferrymen constantly at their post, and the boat is carried over by means of a good and strong rope.[20]

Similar postings for travel down the Southwest Trail between Little Rock and Fulton emphasized the availability of supplies along the route.

> New Establishment, At Washita Crossings
>
> The subscribers, having purchased the property at the Washita Crossings including the Ferry, on the main Military road leading from Little Rock, through Washington, Hempstead county, to Fulton, on Red River, would respectfully inform the citizens of Hot Spring, Clark, and adjoining counties, and the public generally, that they have established themselves there for the purpose of selling goods, under the style of L. Gibson & Co., and have just received and are now opening a large and very special assortment of fresh dry goods, groceries, liquors, &c., which they will sell very low for cash, or in exchange for cotton, peltries, furs, and the usual products of the country.

They invite purchasers to call and examine their stock, as they flatter themselves until they can be suited.

L. Gibson

Wm. R. Gibson

They will keep up the ferry, where travelers and others will be ferried over as safely, and at as low rates at any other ferry on the river.[21]

Although ferries on the Southwest Trail were well maintained, the crossings on Trammel's Trace were not nearly as well equipped or manned. The Sulphur River bottoms were boggy and wide, making it difficult to cross even in dry weather. This obstacle made a ferry crossing at the Sulphur an early priority of the First Congress of the Republic of Texas. In their first session, they offered a half league of land to anyone who would build a ferry crossing at the Sulphur. Mark Epperson was already there, and with this act the operation of Epperson's Ferry on Trammel's Trace was sanctioned and rewarded in April 1837.[22]

Two ferry crossings were available on the Sulphur River, one on each of the branches of Trammel's Trace north of the Barcroft (or Barecroft) headright survey in Cass County. On the branch of the road to Fulton from his cabin up on the bluff above the crossing, Mark Epperson was able to monitor approaching customers. On the Spanish Trace toward Pecan Point and Jonesborough, Stephenson's Ferry crossed the Sulphur River at the present northwestern corner of Cass County. This ferry was established by Joseph A. Stephenson in about 1838 and remained in operation until about 1910.

The crossing of Trammel's Trace at the Sabine River was not served by ferry until 1837, when Francis Ramsdale and his family settled there. Ramsdale operated the ferry between 1837 and 1841.[23] Like other ferry crossings, Ramsdale's Ferry was established near the existing road, where crossings had been made for centuries at natural shoals or shallow water.

Josiah Gregg, a diarist and frontiersman, traveled up Trammel's Trace from Nacogdoches to Arkansas in the winter of 1841, documenting his crossing of the Sabine River. The weather made for a chilly river excursion. A light snow fell on the group only a couple of days after crossing at Ramsdale's Ferry on Christmas Day: "Traveled on what is termed Trammel's Trace, between 35 and 40 miles to Sabine River, crossing at Ramsay's [Ramsdale's Ferry] but it is now not half leg deep,—so forded easily. Its valleys here are marshy and low, and show signs of overflow to the height in many places of ten or twelve feet."[24]

As a consequence of regular flooding along the Sabine River bottom and the muddy thickets on either side of the river at Trammel's Trace, an alternate river crossing developed upstream from Ramsdale's. South of Marshall, a branch of Trammel's Trace separated from the old trail and circled west through Walling's Ferry, ran south through Harmony Hill, and rejoined the original route of Trammel's Trace southwest of Tatum, near the home of Daniel Martin, one of the earliest settlers of Panola County.

Walling's Ferry was not formally established as a town site until 1844, but it was an emerging commercial center for the region. In that same year, Enoch Hays built a two-story, eight-room log tavern and hotel. John Walling moved to the area during the early 1830s and started a ferry operation licensed by the Mexican government before 1836.[25] Walling's Ferry quickly surpassed Ramsdale's as a point of commercial interest, although travelers on horseback or on foot continued to use Ramsdale's to shorten their journey.

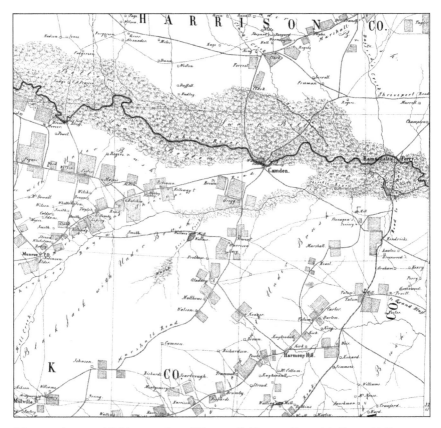

Many roads around Sabine crossing of Trammel's Trace at Ramsdale Ferry, 1863.

In July 1844, John Walling and Larkin Caison were granted a license by the newly formed Rusk County to continue operating their ferry over the Sabine. Fees were set by the county commissioners. Road wagons were loaded and crossed for one dollar, smaller wagons for fifty cents, and a man and his horse crossed for only twenty-five cents. Walling's Ferry found commercial success both as a ferry operation and as a town by crossing livestock. Cattle, hogs, and sheep were three cents each, with footmen and loose horses at six and one-quarter cents each.[26]

Newly formed county governments quickly focused attention on the public interest in roads and ferries. Quite often the first official acts of the counties were to establish responsibilities for roads and appoint overseers for each section. Similar interests with regard to ferry crossings led to a January 19, 1841, "Act Organizing Justices' Courts, and Defining the Powers and Jurisdiction of the Same."[27] The act required ferry owners to uphold specific requirements not only for fees but for the condition of the roads leading to the crossing. The banks on either side had to be kept in good repair, and the slope was not to exceed two feet in rise for each rod in length (approximately sixteen feet). The same act required the chief justice of the county courts to maintain signs at the crossings of all public roads with written directions to the towns or villages to which the roads led.

Trammel's Trace started as an Indian path, used only by those who knew its path well or could discern its way through the wilderness. As its use increased, blazes cut on trees with the blade of an axe allowed those who may have only traveled its length one time on their trip to Texas to find their way. Only five years after the new Republic of Texas formed, the law required road signs along this ancient trail. The transition of Trammel's Trace from Indian trail to public road was virtually complete.

Southwest Trail

The Southwest Trail across Arkansas had a head start on the development of roads in Texas. Its evolution from muddy horse trail to public road continued into the late 1830s and early 1840s. The role of the old military road expanded from sole means of transportation to a feeder for other forms of travel by stagecoach or steamboat. Stage service down the Southwest Trail was established in January 1838. Three times a week, a four-horse stagecoach left Little Rock, passed through Hot Springs, and continued on to Washington, Arkansas, just fourteen miles north of Fulton and Trammel's

Trace.[28] Primarily established for carrying the mail, the stagecoach was also a ready conveyance for any passengers bound for Texas.

The commercial value of property along the Southwest Trail also increased with the numbers of travelers. An advertisement offering to sell a house with stables and storehouses pointed out the value of this stand along the old military road, near a large Cherokee settlement: "all the necessary buildings for carrying on a large mercantile business, and entertainment. The store-house is large and well-finished, and has a lumber-house adjoining. The dwelling is comfortable, and in good repair; and the stables have attached to them good inclosures, and a large grass lot. On the premises is a first rate spring of excellent water, and a large garden."[29]

Commercial interests in trade and travel exceeded the government's ability to keep up with development of the roads. Following a trend in the eastern United States, the Arkansas legislature passed laws allowing the private development of road companies. These early, improved highways were called plank roads or, more often, corduroy roads. They were rough road beds made from whole logs laid perpendicular to the direction of travel. They were hard on both wagons and their passengers but lent some durability and elevation to areas where the road wore quickly or rutted due to the wetness of the ground. Sometimes they were covered with dirt or gravel to ease the passage. Even if they were covered, their rough nature gave corduroy roads a bad name. By the time the Fulton area was mapped prior to the Civil War, the road from Washington to Fulton had sections labeled as a corduroy road. Later maps indicate a corduroy road in a section of the Washington to Fulton road, but only in an area where erosion quickly wore down the trail as it descended a hill.

Edward Cross, a lawyer, judge, and legislator, was instrumental in the passage of a bill on March 3, 1838, to organize and incorporate an enterprising venture called the Washington and Little Rock Turnpike Road Company. The purpose of the company was to build a plank road beginning at Fulton, past Columbus, Washington, and Benton to Little Rock. Had it been built, it would have been for the financial benefit of merchants like Cross, who were eager to develop Fulton into an important river port. It was probably not coincidental to his proposed legislation that Edward Cross was part owner of the gristmill just outside of Fulton and of a substantial part of the property for sale there. There was no indication a commercial plank road ever materialized, although projects of this type did have some commercial success in the more populous northeastern United States.

Hopes for Fulton's importance soared on March 13, 1838; two steamboats passed through the Great Raft and arrived loaded with merchandise and supplies. Fulton was for a time the central point for trade in hides, cotton, lumber, and corn for export to the eastern United States down the Red River and the Mississippi to New Orleans. A market house, school, church, and cemetery were laid out on the Fulton town plot by 1839. Large warehouses and wharves built by Block Brothers lined the waterfront, along with facilities for wool and cotton production. Edward Cross's disputes surrounding land claims in Fulton were finished by 1840, and he again glowingly advertised the sale of lots in the *Arkansas Gazette*.

In spite of the seeming modernization of every form of transportation and the transformation of trails to turnpikes, not every section of roadway got the same attention. Transportation realities were still far from transportation dreams. In one traveler's journey down the Southwest Trail in 1844, he frequently came upon trees that had fallen across the road and were left for years. The lack of responsibility for the roads was something new for those more accustomed to the road maintenance back east. When travelers came upon such an obstacle in Arkansas, they simply made a new trail around the blockage. If something blocked the new trail, yet another turnout was made. Over time the path of the original road was replaced by these side roads that avoided obstacles that could have easily been removed if someone felt responsible. The writer surmised that if he asked those who lived nearby why they did not cut a section out of the first fallen tree, they would probably say that "it is not his business to wait upon travelers."

Trammell and Arkansas Roads

By the second half of the 1830s, Nicholas Trammell settled into his life in Arkansas—keeping a tavern, operating a gambling house, and racing horses since moving to the Washington-Ecore Fabre Road around 1833.[30] He continued all of these business activities for the rest of his life. His sons picked up the same traits and continued these ways of making a living as Nicholas Trammell grew older. Trammell became a notable part of the region where he lived as a result of both the prominence of his tavern's location and his role in developing roads—especially the roads which led past his tavern door, to and from places where the wealthy lived or traveled.

Ecore Fabre was a riverfront trading center to the east of Trammell's tavern that originated as a rendezvous point for French traders on the Ouachita

River. When the road to Washington opened Ecore Fabre to western Arkansas, it became a key shipping point and trade increased. Jacob Barkman, a Trammell cohort, carried on trade from there to Monroe, Louisiana, in dugouts filled with pelts and oil. He returned with lead, powder, whiskey, cotton cards, and Merrimac arrow points for trade.[31]

Ecore Fabre was renamed Camden by migrating South Carolinians and incorporated in 1846. Its location on a high bluff on the west side of the Ouachita River placed it well above the marshy floodplains in other stretches of the river. Its strategic location and accessibility by the river made it the center of trade for a hundred-mile radius on the Arkansas side of the river. Camden had two churches, three schools, a printer, and five mercantile shops by 1848. Hotels and boarding houses served travelers. Unlike Washington, farther to the less-civilized west, Camden had only two licensed dram shops (taverns) that were surprisingly reported as "doing a very dull business."[32]

To make Trammell's tavern more accessible to those wealthy residents, Nicholas Trammell and Robert Carrington were appointed to run a road from Spring Hill to the tavern. They reported the route at the spring term of the circuit court in 1837, and the court appointed George J. Robbison as overseer of the new road. Perhaps to accommodate the anticipated wagon traffic, the commissioners ordered the road be cleared at least twenty feet wide, four feet more than the usual specifications. In earlier times, specifications for road construction meant little to people who rode the ancient trail through the woods, like Trammell in his smuggling days. In those days, a trail was good as long as it was used and cleared by regular traffic. Specifications for public roads became important when local governments were put in place. Good roads had to be cleared of all trees and brush and any growth that could snag someone on horseback. Any stumps in the road, and there were many, had to be lower than twelve inches in height.

More roads from Spring Hill were opened in 1838. Overseers were appointed to open and maintain a road from Spring Hill to intersect the road leading from Nicholas Trammell's place to Greenville in Clark County at or near William Clark's mill. Clark's Mill was about eleven miles east of Washington, not far from Trammell's tavern.[33] In 1838, Nicholas Trammell was instructed to mark out a road to "intersect the Clark County road at the Little Missouri River," just under twenty miles northeast of Washington. Trammell's personal history, his selection of land at a key trading crossing, and his commercial interests put him right in the middle of the development of early roads in Hempstead County, key roads leading past his tavern.

Another road-making assignment given to Trammell was particularly interesting in light of the focus of the court's order. In a Hempstead County action in 1838, Trammell was instructed to change the road leading from Washington to Nick Trammell's tavern so as to run *around* the plantation of Robert Carrington. Carrington owned almost two thousand acres in Hempstead and Miller counties. One one-thousand-eight-hundred-acre tract was along the road toward Trammell's, nine miles east of Washington. Carrington also owned a smaller plot nearer. Carrington was the son of a judge who moved to the Spring Hill region along with many other wealthy planters from Virginia. The fertile lands along the Red River attracted the planters, who brought slaves and their knowledge of plantation operations with them to Arkansas. The community they formed became a place where they could socialize and educate their children. Some of the wealthiest and most influential people in early Arkansas lived around Spring Hill. It is not surprising that Nicholas Trammell was active in opening a road that allowed them access to his gambling house and tavern.

What made the request to reroute the Washington to Ecore Fabre Road around Carrington's plantation even more interesting was Carrington's history in Virginia with a similar situation. Before Carrington moved to Arkansas in about 1832, he was involved in a highly charged disagreement with a man named John Randolph. Randolph was the son of a rich tobacco farmer in Virginia and served the state in both houses of Congress. Randolph and Carrington were neighbors in Virginia. Randolph filed suit against Carrington for trespassing, alleging that Carrington had plowed and planted corn in a road Randolph used to gain access to his holdings. Carrington counter-sued and a neighborly feud began between two men used to having their own way. Randolph was reportedly becoming somewhat deranged, making the disagreement more volatile. When Carrington sought an alternate outlet from his estate over a tract of land Randolph owned, three men were ordered by the court to view the road. They arrived only to find Randolph had posted a notice on the gatepost with the names of all who could pass along the road. The viewers read it carefully, comparing it to the names of every man and woman in the neighborhood. Only the name of Robert Carrington was omitted from the list.[34] Perhaps the memory of that encounter led Carrington to request a rerouting of the road around his plantation so that he might avoid any others so uncooperative as Mr. Randolph.

Trammell's strategic location along important and well-used roads through southwest Arkansas was a key part of his success as a trader and as an operator of a gambling establishment. When he settled in Hempstead County, it was likely that his broader trading ventures were reduced in scope and that gambling, horseracing, and farming took their place. Given that his family continued to identify themselves as traders for many more years, it is unlikely that Trammell gave up the opportunity for trading completely. Since his tendencies for trading lingered, Trammell was no doubt intrigued when word of a new route along the Red River to the riches of Mexico was circulated in August 1839.

Trade with Mexico had commenced for years from St. Louis to Santa Fe and then south toward Mexico and Chihuahua City. In 1838 and 1840, a group of traders from both the United States and Mexico blazed a shortcut from Chihuahua to a trail across Texas. Their route, the Chihuahua Trail, pushed northeast from Mexico toward the Red River at Pecan Point and along the south bank of the Red River to Fulton. The old route from St. Louis, heading west and then turning south at a ninety-degree angle, was 1,350 miles west to Santa Fe and then 650 miles south to Chihuahua. By cutting that right angle diagonally from Chihuahua to Pecan Point, bypassing Santa Fe, they reduced the journey by 1,200 miles.[35] This seemed to be a significant business advantage to the group blazing the trail. They made the first trip without any attacks by Indians or encounters with other obstacles. The group returned to Fulton in April 1840, with eighty wagons full of merchandise to take back to Mexico. Oddly, they were also accompanied by a company of equestrian performers on their way to entertain in Mexico. For five months the wagons endured the heavy river bottoms and bogs south of the Red River, made worse by above-average spring rains. Even after they reached the drier prairies, their trip took three more months of travel.[36] The time and expense of the trip made it a losing venture. No further trade commenced, and the only outcome was the creation of a new roadway worn by the wheels of eighty loaded wagons.[37]

Cherokee Expulsion

As President of the new Republic of Texas, Sam Houston continued his attempts to help the civilized Cherokee tribes in eastern Texas achieve their goals of securing land they could call their own. When the land offices across Texas opened in February 1838, the chances of that ever happening were lost in the rush to Texas for land. When Mirabeau B. Lamar was inaugurated as new president of Texas in December of that same year, Cherokee

Indians in eastern Texas not only lost their land, but they were about to be completely removed from Texas.

For the Cherokee in the large village not far from Trammel's Trace, north of Nacogdoches, 1839 was a time of transition and upheaval. By that summer, they lost patience with the promises for land. A survey of the land promised had been completed, but the treaty was not acted upon by the Texas Congress. Alexander Horton, who served as *aide d'camp* for Sam Houston during the Texas Revolution, was called out to monitor the growing restlessness. He was ordered to "guard the Sabine river from Logansport to Trammel's Trace and to prevent the Cherokees from being supplied with arms and ammunition by other Indians and Mexicans" from the eastern and southern boundaries of Texas."[38] News of the Indian troubles in eastern Texas traveled up Trammel's Trace into Arkansas, reaching Fulton as a group of travelers prepared to cross the Red River. As a result of the growing conflict, they were advised to avoid Trammel's Trace and head down the eastern bank of the Red River to Natchitoches and cross into Texas by way of the Sabine, advice they surely followed.

National Road of the Republic of Texas

Roads in Texas were generally behind those in the United States in terms of their development, but the new nation of Texas made ambitious efforts to catch up, given its limited resources. The historic El Camino Real was the focus of an 1839 bill by the Texas Republic calling for road improvement.[39] For north-south travel through Central Texas, the Central National Road of the Republic of Texas was authorized by the legislature in 1844. Five men were commissioned to survey and establish the road. This portion of a larger road network for Texas was to begin on the Trinity River in present Dallas County and run north to the south bank of the Red River. Its terminus there was to be opposite a key location established earlier in Texas history—the mouth of the Kiamichi River, six miles upriver from Jonesborough. With earlier roads connecting along the Ridge Road, Trammel's Trace, and the Southwest Trail, Texas' Central National Road effectively created an international route between St. Louis and San Antonio.[40]

The village of Kiomitia that developed at that northernmost point on the Central National Road was the former location of Wright's Landing. Wright's Landing was generally accepted as the upper limit of navigation on the Red River. What seemed a lonely spot on the map for anyone not familiar with the history of the Red River settlements was in reality the location of a strategic confluence of events only twenty to thirty years before the Central National Road was surveyed. The road to Fort Towson, other roads from the east, and the history of settlement naturally led the surveyors to this point. They knew the history of the region well. Three of the five surveyors lived near the Red River. Until the Central National Road was developed, the residents along the Red River, with their roads running east and north, were really more a part of the economy of the United States than the Republic of Texas. The only road running south from the Red River was Trammel's Trace to Nacogdoches.[41] This new central road would help align citizens along the northern borders of Texas with the new Republic.

Trace Still Dangerous—Bloody Prairie

During the early years of the Texas Republic, many who traveled Trammel's Trace across what is now Harrison County, Texas, described the beauty of eastern Texas. Beyond the brushy riverbanks, there was little undergrowth beneath the tall, straight timber. Small prairies devoid of trees were covered with lush grasses that stayed green for much of the year.[42] There were springs about ten miles north of the Sabine River crossing of Trammel's Trace. Indians who believed the springs had medicinal benefits were attracted there for centuries. Caddo burial mounds dot the nearby woods and fields. Shaded by a grove of pines, the springs bubbled up in eighteen locations within a small area. The springs seemed an inviting place to settle. The close proximity to Trammel's Trace meant the springs were frequented by many who traveled the trail, some of whom were not always a desirable lot. Although the landscape around Rosborough Springs seemed peaceful and serene, the nature of the people who remained after the settlement of the former Neutral Ground made it dangerous. The area just south of Rosborough Springs was known as Bloody Prairie—one hundred acres of open land where many misdeeds took place, earning the name as a murderous field of danger.

On September 30, 1837, two men from Missouri, Loveland and Smith, were riding on horseback down Trammel's Trace. About five miles north of the Sabine River, they discovered an unusual sight. Although there was no one around, a fire was still burning in the middle of the road. More practical than curious, the men decided to take advantage of the fire and cook some breakfast. But after they walked into the woods to gather their horses, they discovered the body of a man recently killed. From the tracks they could tell two men on horseback had left the scene. They discussed the gravity of the matter and decided to follow and apprehend the perpetrators. They found the perpetrators' trail and overtook the two killers thirty miles south down Trammel's Trace. The two Missourians commanded the men to stop. Subduing the killers, they took them back to the place where the body was found.

One of the murderers, a man named Carson, said his companion, Mr. Quarles, shot the man dead and dragged him into the woods. What Loveland and Smith soon learned was that there was a second dead man in the woods, beaten with the butt of a rifle. Carson pointed the finger at Quarles for both killings, so Loveland and Smith put a noose around Quarles's neck and threatened to hang him until he confessed and said where the victim's gun could be found. After the facts were clear, the captors took their two prisoners down Trammel's Trace to Nacogdoches, where they awaited trial in the criminal court. Quarles was hanged in San Augustine, but Carson escaped across the border to the United States. The victims, McFarland and Turner, were likely buried where they were found, five miles north of the Sabine River on the edge of Trammel's Trace.[43]

Bloody Prairie earned its name from more than one occasion. John Roland built a cabin on the prairie within sight of the springs but abandoned it in 1836 when a nearby family was murdered by Indians. A man named Tidwell was also killed and scalped in front of his wife and their three small children, who were held hostage in the band's village far to the west near the Brazos River. Delaware and Chickasaw traders somehow notified the family of their plight. Mrs. Tidwell's brother, Richard Blanton from Lafayette County, Arkansas, traveled to Texas and recovered them from the Indians. While Mrs. Tidwell was held captive, she saw a young white boy named Thomas Pierce, seven or eight years old, among the Chickasaws. His family had also been murdered on Trammel's Trace two or three years prior.

A fork in Trammel's Trace at the Bloody Prairie north of the Sabine offered travelers a choice. Continuing west led them to Walling's Ferry and its growing commercial activity. That route continued through Harmony Hill and rejoined Trammel's Trace south of Martin Creek. Turning south at Bloody Prairie led them to Ramsdale's Ferry, over five miles of rolling country to the Sabine. By either branch, Trammel's Trace was the preferred and recommended route south to Nacogdoches in 1846 in spite of the dangers. "The road is remarkably good, and for the most part level, and easily traveled. It is the most agreeable route, and far better than the road from Natchitoches to the Sabine [El Camino Real]. Indeed the whole route from Shreveport, either by Walling's, Brewster's, Ramsdale's, or Logan's ferry, on the Sabine is decidedly better for emigrants who are going westward... The distance is shorter, the road better, plain, dry, and gently rolling, and far less fatiguing to man and beast than either of the old routes before mentioned, farther south."[44]

The trail may have been more favorable, but the region served by Trammel's Trace was socially unstable. In a section of the former Neutral Ground served by Trammel's Trace, Anglos were fighting each other over land between 1839 and 1844 in a five-year battle called the Regulator-Moderator War. Primarily involving the residents of Harrison, Shelby, and surrounding counties, this conflict began with fraudulent land certificates evolving into theft and murder for no apparent reason other than the sides one chose. One of the most notable events during this time was on March 2, 1842, when Capt. William Pinckney Rose gunned down Robert Potter, a leader of the Moderator faction and the former secretary of the navy for the Republic of Texas.

Other than its use for men at war over land, Trammel's Trace was still a thoroughfare for routine travel. When Indian Commissioner G. W. Hill traveled from the Brazos to the Red River in June 1843, he noted that ferry service was available at every river crossing along Trammel's Trace. Ramsdale's Ferry at the Sabine charged $1.00 going north and $1.50 coming south. Pugh's Ferry at the Big Cypress near Jefferson and Epperson's Ferry across the Sulphur River were each one dollar. Langford's Ferry across the Little Cypress was the most expensive at four dollars. Besides the availability of ferriage, Hill's expense report indicated that he had little problem finding lodging and meals along the route, an indication of the rapid growth and immigration across the area.[45]

Family Issues—Marriage, Debt, Death

In the years after Trammell established himself along the Washington to Ecore Fabre road, his family life and business life were far more settled than during his turbulent years in Texas. Trammell was fifty-three years old when he moved to Hempstead County. His years as a trader and smuggler were over, and his gambling and tavern operations did not rely on his travel for hundreds of miles across perilous territory.[46]

Beginning in 1833, Trammell lived in southwestern Arkansas on 160 acres in the southeast corner of Section 25, Township 12 south, Range 23 west. He purchased an additional eighty acres adjacent to this farm in 1841. Curiously, that land was previously owned by Isom Parmer when he left Arkansas. Parmer's father, Martin Parmer, was the man who killed Mote Askey.[47] Tax records show Trammell owned seven horses, but their taxable value was only $350. Either they weren't the finest stock of racehorses or he had managed to deflate their value for tax purposes. Four mules and twenty head of cattle rounded out his stock. Trammell also owned four slaves in 1841, likely to work both the farm and the tavern. Seven people were in his household, including Nicholas and his wife, Sarah.[48]

His sons and daughter had come of age and developed their own lives as well. Trammell's oldest son, Nathaniel, moved between Lost Prairie, the flat, fertile land where acres of canebrakes were replaced by cotton, and ports near the Mississippi River. Nat had a number of children in Arkansas and married a second time after his first wife's death. Another son, Phillip, also owned land in Lost Prairie and enough slaves to operate a small plantation. The land Phillip farmed was the same site his father first cut out of the canebrakes of Lost Prairie after his retreat from Texas. Phillip ran his own tavern along the road from Dooley's Ferry to Texas.

Nicholas Trammell's son Robert may have been a partner in Phillip's tavern, and there are documents to verify his movements. Robert had four children, two born in Arkansas and two born later in Texas. After the Texas Revolution, Robert remained south of Fulton and west of the Red River in the expansive Red River County, which covered most of what is now northeast Texas. Robert Trammell's headright survey is now in present Marion County, Texas. In 1838, Robert signed a petition to the Texas legislature requesting the formation of new counties along the Red River.[49] In a request that opens some interesting questions without providing answers, Robert petitioned the court in August 1838 to be appointed administrator of the estate for Mote Askey. Twelve years after Mote was killed, far away in Nacogdoches, why would Robert petition to represent the estate? The strongest possibility is that he was doing so on behalf of Mote's son, Nicholas Harrison Askey. Nicholas Trammell had raised Harrison, who by the time of this request was approaching eighteen years of age.[50] Perhaps Harrison was living with Robert Trammell instead of Nicholas at the time. The court ordered that Robert's request be continued to the next term of the court, but no further action has been found in the records, and there is no information about what might have remained in Mote's estate.

Another interesting transaction for Robert came through Red River County on September 16, 1840. Robert Trammell had a substantial debt he could not repay to two traders named Carson and Hall. Jesse H. Hall was a partner in the plantation-scale planting enterprise known as Carson, Butler, and Hall, so Robert likely had a debt related to farming operations.[51] His father helped him out of the jam and paid off the debt. In exchange, Robert transferred slaves and other property to Nicholas.

> Know by all these present, that I, Robert Trammell, of the Republic of Texas and Red River County, have this day bargained and sold to Nicholas Trammell of the County of Hempstead and the State of Arkansas the following slaves:
>
> Sally Touch, about 23 years of age, and her two children one four years old and one one-year-old, and a man, Bob, lame in the foot. Five years from the 23rd of April, also half of the crop of cotton and corn now growing on the plantation where I now reside and seven bales of cotton at Halls Gin, two mares and three heads of horses—3 cows and also 3 work steers—20 head of hogs for valuable consideration of eighteen hundred and fifty dollars, which the said Nicholas Trammell is to pay the estate of Carson and Hall for the said Robert Trammell. . . .[52]

Nicholas Trammell was apparently not the type to make a gift of any kind, certainly not for such a large amount of money. Even though Robert may have gotten in over his head, his father was still there to help, although with his own conditions.

Although arrangements to pay the debt had been made, both Robert Trammell and Phillip Trammell were summoned by the District Court of the Republic of Texas in Bowie County on October 20, 1841, for a different matter. They were to appear in court the following March to resolve a suit by Charles Collum, administrator of J. H. Hall's estate.[53]

Nick's son Henry, born in 1817, was in many ways the most enigmatic of the children of Nicholas and Sarah Trammell. In May 1842, at the age of twenty-four, Henry Trammell married eighteen-year-old Julia Ann Thurmond in Hempstead County. As a consequence of many later legal actions involving Henry, details of his activities are well documented. That he lived apart from his father and siblings at times and was in repeated legal disputes with his brothers presents questions for which answers do not remain. More about Henry will be presented in the next chapter.

Nicholas Trammell's daughter, Nancy, left home at the age of twenty to start her own family, marrying Green Baker in Hempstead County, Arkansas, on September 19, 1839. The proud father of the bride gave his daughter a wedding gift in a way that protected the property he granted. Nicholas Trammell was ever the businessman. His troubles in the past and experience in legal matters taught him to look ahead: "I, Nicholas Trammell, in consideration for the natural affection I have for my daughter, Nancy, now the wife of Green Baker now residing in the county of Lafayette in said state. . . in consideration of the sum of one dollar, good and lawful money in hand paid by Philip Trammel, I do hereby give, bequeath, grant, and sell to said Phillip Trammell in trust a certain negro slave named Mary and her boy child for use and benefit for my beloved daughter, Nancy, and heirs of her body for the duration of her natural life, and her children after her death, if there are any. In case there are none, said slaves shall revert to the estate for the benefit of the heirs of the said Nicholas Trammell."[54]

By transferring ownership of Mary to his son Phillip for Nancy's benefit, Trammell ensured that ownership of the slaves would not transfer to Nancy's husband in the event of her death. All of Trammell's dealings with crafty attorneys and judges had not passed without his learning something about how the law worked.

Life seemed regular and predictable for Trammell. Then, in the crisp, fall weather of late September in 1841, Trammell's wife, Sarah, died.[55] They had been through much together, moving from Kentucky to Missouri, from Arkansas to Texas, and then back to Arkansas. With Trammell's trade crossing states and nations, Sarah must have been a touchstone for her five children and for Nicholas Trammell as well. She raised a family while Trammell was away on trading ventures. Her good standing with the Indians may have prevented the massacre of her family at the Trinity River crossing. And she endured the difficulties of being chased out of Texas.

One hundred years after her death, some of the myth of Nicholas Trammell was built upon stories that hinted that perhaps his wife did not die a natural death. Storytellers said his wife was buried at night in an ancient burial ground close to the tavern and that an excavation found a single set of female bones there. There is no record of her cause of death, no burial marker, but there is a curious juxtaposition. On the same day that his wife died, Nicholas Trammell recorded a transfer of property in a deed of trust. The record shows Nicholas Trammell transferred ownership of a "d____d [damned] nigger" to his son, Phillip.[56]

Did one of his slaves accidentally kill his wife, Sarah? Was one of them somehow responsible for her death, and was the only way to dispel his anger and sadness to get the slave out of his sight? Rather than the cold-blooded response of the rumored murderer and outlaw that myth and legend painted him, it appears that Nicholas Trammell tempered his sadness and anger with an act of conscious tolerance. Add to this Trammel's involvement in the "rescue" of the chained slave in Nacogdoches, and a different image of Nicholas Trammell emerges from the facts.

After living along the Washington–Ecore Fabre Road for ten years, next to the creek that would ultimately take on his name, Nicholas Trammell finally received the federal land patent for his new homestead on March 1, 1843.[57] Ten years of waiting allowed Nicholas Trammell and his family to recover from the chaos in Nacogdoches and his expulsion from land they hoped to settle in the Texas territory. From a safe distance, Trammell watched the events in Texas following his inglorious exit from that disputed territory. Finally, he was reaping the benefits of his stability and the intricacies of public land policy in the United States. With a few horses and cows, a steady stream of travelers passing his tavern who were willing to be relieved of some money, and a few slaves to tend to things, Nicholas Trammell transformed himself from clandestine trader to settled farmer and businessman.

Apparently Trammell's life was not complete without a wife. Two years after his first wife's death, he married a second time. This time, Trammell married a woman not much older than his own children. Mary Sadberry became his wife on May 21, 1843. Jacob Whitesides, an elder of the Methodist Church, performed the marriage ceremony.[58] Nicholas was sixty-three years old when he remarried; his young wife was only thirty-one. The Sadberry name was found across southwestern Arkansas. Mary was likely related to Marvin T. Sadberry, who owned land near Spring Hill about a mile off the road to Dooley's Ferry. Interestingly, Nicholas Trammell named the two children of this marriage for himself and for his first wife. A son, Nicholas Trammell, was born in 1844, and in 1845, Mary gave birth to Trammell's seventh child and second daughter, Sarah Jane.

Trammell's age and marital status did not relieve him of his obligations to maintain roads in the area. At the October 1843 term of the Hempstead County court, Trammell was fined $1.50 for failing to work on a road in Missouri Township for one day, violating a previous order of the court.[59]

Along with the happiness of marriage, life in Arkansas for the Trammells included loss as well. Phillip Trammell wrote his will on February 19, 1844, and on February 20 he died in Lafayette County, Arkansas.[60] The cause of his death is not known, but it appears to have been unexpected and sudden. His father was a witness to the will passing the entirety of Phillip's $36,000 estate to his son, also named Nicholas, and naming his brother Henry as executor. Phillip's scrawled, pained signature directed that his farm and slaves be kept together and that the proceeds of the estate be used to educate his son at one of the best colleges in the United States. Young Nicholas was eight or ten years old at the time his father died. The estate was sizable, with thirty slaves of various ages among the property.

An unusual aspect of Phillip's will directed that Henry would not be required to pay the bond normally mandated for an executor. In spite of the will, only a few months later, in May 1844, an order was issued requiring Henry to post bond in an amount two times the estate or his letter of administration for the estate would be revoked. However, Henry continued to administer the estate and to care for Phillip's son, Nicholas. Henry's suspect actions led to a lingering court battle among the family.

If any of the people Nicholas Trammell encountered in his years of trading were in Texas in 1844, they may have read in a Houston newspaper that the old smuggler had been shot. Many folks probably wondered how he escaped that fate for so long. The report was not true. Nicholas Trammell, a new father at age sixty-four, was alive and well in Hempstead County, Arkansas.

Newspapers of the period were notorious for embellishing the facts, and the way this story evolved demonstrates how misinformation was propagated. An article in the *Arkansas Gazette* on March 27, 1844, described how an apparently unrelated John Trammell from Hot Springs County was shot by a man he was holding captive, awaiting trial at the circuit court. In an unguarded moment, the criminal seized John Trammell's gun, shot him, and killed him instantly.[61] That was the end of the Arkansas version of the story. When the same story was repeated in the *Houston Morning Star* death notices on April 18, 1844, one additional sentence of editorial comment was added.

> Mr. TRAMMEL, we believe, is the person who first marked out the road, known as Trammel's Trace in Eastern Texas.[62]

The rumors in Texas of Nicholas Trammell's death probably made it back to him in Arkansas. It would not be the only time his death notice appeared in a Texas newspaper.

9

1845–1856
The Old Smuggler Retires

'Tis but a blot against the sky tonight
But once it was an inn—a pretty sight
To travelers with money all there won,
Who grew weary of the wagon trail alone.

—ALTA HONEA[1]

Nothing in Texas would ever be the same after it became part of the United States. In 1836, the total population of Texas consisted of approximately thirty thousand Anglos, four thousand Mexicans, and five thousand slaves. By 1847, there were one hundred thousand Anglos and forty thousand slaves.[2] When Anson Jones relinquished his post as the last president of the Republic of Texas in February 1846, he proclaimed the "final act in this great drama is now performed, the Republic of Texas is no more."[3] Texas' statehood in 1845 was celebrated by most of those who lived in the republic. It was the logical conclusion to all the previous years of filibustering, insurgence, and revolution. Although statehood seemed to be preordained since the earliest rebellions in Nacogdoches, it took almost ten years after Texas' independence from Mexico for that transition to occur.

Mexico did not celebrate. Instead, Mexico immediately broke off diplomatic relations with the United States over the issue of statehood. Yet another boundary dispute over the Texas border with Mexico ensued, and again, Gen. Santa Ana was the focal point. The United States invaded Mexico, and war was officially declared on May 13, 1846, after the first battles had already been fought on the Rio Grande. Mexico did not invade the expanding nation to its north and start a war. There was war because the United States saw an opportunity for expansion. By the time the war ended two years later, the United States had gained most of what is now the southwestern United States and Mexican territory decreased by half.

War with Mexico Passes through Arkansas

When President Polk called the nation to war in 1846, Washington, Arkansas, became a rendezvous point for soldiers assembling from Arkansas, Tennessee, and Missouri. Ten companies from Arkansas alone mustered at Washington. That town's strategic crossroads and proximity to the Red River in Fulton made it an ideal place from which to stage troops heading to battle through Texas and along Trammel's Trace. Washington's prevalence of grog shops certainly would have had a brisk business from the soldiers. At the time, Washington had twelve taverns but only one church.[4]

Military leaders planned for troops to cross the Red River at Fulton and proceed southwest toward San Antonio following Trammel's Trace or to take the Fulton-Natchitoches road toward the El Camino Real. When supplies did not arrive in Fulton, one regiment changed its plans to receive supplies at Robbin's Ferry on the Trinity River, the same crossing once occupied by Nicholas Trammell. The commanders fully met the challenges of river crossings with men and supplies. Using three inferior boats, over eight hundred men and horses with forty wagons of supplies crossed the Red River in only a day.

One of the regiments that passed through Washington was later captured and held prisoner at Encarnacion in Mexico. Following the release of the prisoners, an anonymous soldier wrote of his travel across Arkansas and into Texas, providing a detailed description of their guide into Texas—none other than Nicholas Trammell.[5]

The Kentucky Cavalry, led by Colonel Marshall, arrived in Washington during the summer months of 1846. Their journey from Kentucky had been long and difficult. The soldier's account said they were plagued by snakes, tarantulas, and centipedes all along the way. When the regiment arrived in Washington, wagon repairs and horseshoeing required a five-day stopover near the hilltop in Washington. Conflicting accounts of the best route to Texas left Colonel Marshall somewhat perplexed. The upper route through eastern Texas, Trammel's Trace, was reported as best supplied with water but a rough trail and thinly inhabited. The lower road, south along the Red River to Natchitoches, was said to be shorter and more open but badly watered. Watkins Cocke, a resident of Washington, told Colonel Marshall that there was an old man in the county who knew all the roads in Texas and could answer his questions. The man's name was Nicholas Trammell.

Trammell's former problems in Texas were apparently widely known across southwestern Arkansas. The soldier knew Trammell had been a freebooter all his life and that Texas had been the primary scene of his illegalities. The facts of Trammell's expulsion from the ferry crossing on the Trinity were already being magnified and embellished. Everyone said Trammell was run out of Texas as a result of being so "audacious in his enterprise of villainy that the settlers were alarmed for their imperiled firesides."[6] Only twenty years after the well-documented events, Nicholas Trammell's myth was already expanding. The young soldier was impressed with Trammell and provided the only firsthand physical description of the well-known smuggler: "Trammell is about fifty years of age, five feet ten inches in height, of a spare but sinewy frame, having a long thin face and small grey eyes. He is intelligent and shrewd, and very agreeable in conversation."[7]

Although Trammell was actually sixty-six at the time, his physicality belied his age. The soldier's physical description of Trammell was not the most telling aspect of his story. It was his observation of Trammell's demeanor and reputation that captured the personality of this life-long frontiersman and trader. When Cocke introduced Trammell to Colonel Marshall, Cocke made it clear that Trammell "would perform with fidelity and honor, whatever he undertook, but that it was prudent to watch him after he had completed his engagement."[8] How interesting that this 1846 description of Nicholas Trammell as an old man, settled in Arkansas, matched a similar description generally applied to Anglos by Chief Connetoo, one of the Cherokee who accused a younger Nicholas Trammell of horse thievery in 1813: "When a white Man enters my house I treat him in a stile to render him comfort. When he leaves my house he fails not to run off with some of my property."[9]

This ethic of the frontier, that theft was acceptable as long as the victims were caught unawares or tricked out of their property, was a dichotomy consistently displayed by Nicholas Trammell since his public deeds emerged in 1813. Colonel Marshall was willing to take advantage of his knowledge but remained wary of Trammell's presence until his services were no longer needed. Colonel Marshall hired Trammell to guide his regiment from the Red River to Trammell's former home at Robbins Ferry on the Trinity River.

Marshall, two of his soldiers, and Trammell set out to mark the route, blazing trees with an axe cut along the way. Three days passed before Kemp Goodloe, one of the escorts, returned to Washington with word of his agreement that Trammel's Trace would be their route. Goodloe reported that it was over a high, shady terrain through "stately forests of fine trees and along the margins of extended prairies spangled with flowers and glittering in green."[10] Who could resist such a route? Thirty men, half the regiment, were sent ahead after they crossed the Red River to clear the road in front. The other half brought along the wagons through the newly cleared path. Eight miles of dense canebrake faced them on the rich river bottom past Fulton's ferry.

Marshall and Trammell continued ahead of the rest of the regiment. Marshall's mistrust of Trammell grew as they wound their way along paths unknown to the Colonel. The soldier writing of these events reported that Marshall tired of Trammell and sent him back to the regiment. Marshall apparently believed he could do just as well without Trammell. Marshall's naiveté about the terrain resulted in his leading the regiment on torturous reversals of progress. In attempting to avoid the lakes, swamps, and bayous along the route, Marshall's guides often led them in circles. Some of the men declared that they had bought butter and milk at the same cabin for five days in a row while making no more than ten miles progress across the thickly wooded creek bottoms. Part of the problem was that without Trammell's direction, Marshall was left to the advice of the few inhabitants they encountered. Every branch of the trail to a cabin door was a diversion that the inhabitant hoped would become the main road past their home. Word of Marshall's plight apparently spread quickly. Trammell waited patiently with the regiment, knowing that Marshall would have to turn to him again.

Trammell eventually rejoined the soldiers and continued down Trammel's Trace, no doubt reminiscing about his many trips along the old trail in both directions. When the soldiers arrived within fifty miles of Nacogdoches, perhaps at Ramsdale's Ferry, Trammell stopped and determined he would go no further. He explained that because of the reasons surrounding his sudden departure from Texas, there were many in that country that might recognize him and recall his past behavior toward them.

Whether Trammell had been back down that stretch of the Trace since he was chased back to Arkansas twenty years before is unclear. With his business in Arkansas focused on the tavern and on racing horses, it is entirely possible that Trammell had no further use for this part of Texas, leaving any business connections there to his sons or to other traders. He and his sons were spread out all across southern Arkansas and northeastern Texas, and business was good there. A few short years later, however, Trammell overcame his concerns about Texas or submitted to the wishes of his sons. He and his entire family would move back to Texas once again.

Trammel's Trace Evolves

Increasing migration through Arkansas into Texas led to a rapid development of other roadways across the territory. Problems with the routing and construction of the Southwest Trail (the old Military Road) across Arkansas led to its replacement in many sections. A new road advertised in 1847 provided a route advantageous for travelers: "about 12 or 13 miles nearer from Little Rock to Washington, than the old Military Road, and by far the best road. It runs through a fertile, level, and rich country, and as soon as travelers find it out, it will be traveled altogether, and it will bring the right lands of this country into notice at once."[11]

With the end of the war with Mexico in 1848, immigration to Texas grew even more, and the pace of settlement between Nacogdoches and the Red River increased. The new waves of immigrants were from Alabama and Mississippi. They entered Texas at Gaines' Ferry and then moved northward up Trammel's Trace from Nacogdoches. Although newer roads between settlements were taking on more of the traffic than Trammel's Trace, the old trail remained a significant part of the landscape in the years of early statehood for Texas. The inhospitable wilds of Trammel's Trace were being tamed by road builders appointed by local governments. Near Jefferson, Texas, the section of Trammel's Trace across the bottomland between Eli Langford's ferry at the Big Cypress and Frances Pugh's ferry across the Little Cypress Bayou was appropriated as a public highway on January 25, 1848.

Some of the earliest action taken by new county governments was the establishment of roads and ferries. "Marking and clearing" roads by the best route was no longer sufficient instruction for maintaining roads for transportation and trade. Roads were classed according to the specification required for their clearing. First-class roads were thirty feet wide with stumps cut down to six inches high; second-class roads were twenty feet wide. Males between age eighteen and forty-five and male slaves from sixteen to fifty were required to serve in road maintenance crews in their precincts. Ministers, postmasters, ferrymen, and millers were exempted, along with any

Indians who remained. Designation as an overseer for a road meant taking responsibility for clearing trees and brush to maintain it.

An idea for the commercial development of roadways found its way from the eastern United States into the emerging states. Edward Cross remained one of the largest landowners at Fulton, and he continued his efforts to make Fulton a commercial center through legislation and commerce. As a follow-up to the proposal to connect Fulton by a plank road to the north, Cross introduced a bill in the Arkansas legislature in January 1851 to create the Ouachita and Red River Plank Road Company.[12] Plank roads became a popular enterprise across the country, many involving speculators. Some turnpike roads in New York were paying their investors as much as a twenty percent return. Turnpikes were authorized by public authority operating at the expense of private investors and funded by tolls on the users. Payment of the tolls released travelers who lived nearby from maintaining the road themselves.[13] The route specified for this ambitious road across southern Arkansas was along the same roads frequented by Nicholas Trammell, including the one right past the door of his tavern. The road was to be constructed from Camden "in the western direction, or as near as may be, on the best route to or near Fulton on Red River."[14] There is no evidence to suggest it was ever built.

A similar turnpike corporation was organized in Texas to cross one of the more difficult stretches of Trammel's Trace. Mark Epperson, operator of the ferry across the Sulphur River since 1837, was authorized by the Texas legislature to incorporate the Sulphur Fork Turnpike Company in January 1852. Their mandate was to build a plank road across the river bottoms on either side of Epperson's Ferry. Bridges across the sloughs and a road above the normal overflow was a grand vision, but conditions made the task impossible to accomplish. After four years with no road, the legislature amended the act to eliminate Epperson's partners and leave him with sole responsibility. There is no indication Epperson's project ever got underway.

The old smugglers' trail through uninhabited forests in northeast Texas was designated as public highway, and the road from Trammell's tavern toward the Red River was envisioned as paved with logs. Grand ideas did not always match reality, but these visions for the former wilderness were unmistakable indicators that the time for clandestine travel had passed from the realm of Nicholas Trammell and Trammel's Trace.[15] The advancement of civilization into new territories was not merely a movement of people: it was also a transition of culture. Although the wildness of Trammel's Trace was being tamed by conscripted overseers, the reputation of those who had settled along its path continued to be questionable.

One observer noted that the inhabitants of frontier settlements fell into four classes. The first class included the hunter, trapper, and Indian traders arriving far in advance of any civilization. Nicholas Trammell and his kin had been a part of that advance through Kentucky, Tennessee, Missouri, Arkansas, and Texas. Backwoodsmen made up the second class—those who chose to live apart and on their own. They often lived closer to the Indians than to the Anglo settlements, providing a buffer of sorts between the two distinct cultures in the Neutral Ground. The Red River settler at Jonesboro and Pecan Point certainly fit this category. In the parts of Texas and Arkansas through which Trammel's Trace crossed, a third class of inhabitants became predominant. In a place where only hoof prints had been, planters, merchants, horsemen, and farmers were moving in and changing the landscape forever. They made way for the fourth class to come, the inhabitants of the more densely populated towns, most of whom would have little understanding of the difficulties faced by the men and women who settled this country only thirty years before.[16]

The earlier reputation of those living and traveling along Trammel's Trace as bad men and outlaws was well known and broadly repeated. Even after Texas' statehood in 1845, Trammel's Trace was still said to be a place where murderous desperados gathered with others of their kind, ready to engage in any shadowy effort. The reputations of Trammel's Trace and those who chose to live along its forested path were both tainted by their past, as this description of one Harrison County settler implies: "Although his cabin was near Trammel's Trace, and in part of the frontier where many of the setters were men of bad character, yet he was respected and beloved by all. . . ."[17]

A good man living along Trammel's Trace had to qualify his character or be cast among the outlaws based simply on the location of his residence.

Something Stirring in Texas

In 1845, the year Texas became a state, Nicholas Trammell turned sixty-five years of age. He had no doubt outlived many of his contemporaries and outlasted any legal authority that had challenged him since his 1813 accusation by the Cherokee. Twenty years after he was run out of Texas, he had established Trammell-operated ventures from the Mississippi River in southeastern Arkansas to the upper Red River west of Spanish Bluff. Although his primary businesses of farming, tavern keeping, and gambling had become the primary work of his sons, Nicholas Trammell was still the patriarch of his family. Those who knew about Trammell from their firsthand experience with him in the Missouri Territory forty years earlier were mostly gone. Their sons and daughters may have heard tales of Nicholas Trammell, along with the embellishments that enhance such stories. Anyone who could correct the details was probably already dead, and those hearing their stories were more interested in a good tale than an accurate accounting. Nicholas Trammell was a key figure in southwestern Arkansas and northeastern Texas, and his involvement in historic events was well documented. Stories of Trammell the outlaw would only get better over time.

Although his businesses were well established across southern Arkansas and northern Texas along the Red River, after Texas statehood, Trammell's focus shifted once again to opportunities inside that vast territory. Nicholas Trammell's ties to Arkansas dwindled in his old age. The unsettled and ambitious Trammell family planned to return to Texas.

1845—Trouble Begins, Family Lawsuit over Slaves

Land sales and court appearances provide clues to the Trammell family's transition to Texas. As with every move Trammell had made since leaving Logan County, Kentucky, it was never a clean break but more of a resettlement over time. Actions taken by Trammell's son, Henry, gave some indication of when the family began preparing themselves to leave Arkansas behind.

Henry Trammell married Julia Ann Thurmond in May 1842, when he was twenty-five years of age. The Trammells always had their share of legal problems, and Henry's new wife came to the family with her own disputes, which lingered for over twenty years. When Julia's grandfather, Richard Thurmond, died, he left a will granting slaves to his grandchildren. Ownership of the slaves was later questioned by others who claimed an interest. William McDowell Pettit filed suit in the circuit court in Chicot County, Arkansas, on November 12, 1845, naming Julia Thurmond Trammell in the suit. It also named her brother and sisters William Thurmond, Celia Thurmond, and Rebecca Thurmond, along with the husband of their late sister Judith, James Hutchinson.[18]

In the suit, Pettit claimed title to the same slaves Richard Thurmond gave his grandchildren in his will. Since some of the Thurmonds lived out of state, Pettit ran a notice in the *Arkansas Gazette* ordering them to appear in court at Columbia, Arkansas, in Chicot County, at its next term in May 1846. When the date for that court term arrived, a subpoena had not yet been served on Julia Thurmond Trammell because the court found that she was "proven to be a non-resident of this State."[19] Notices appeared in the *Arkansas Gazette* in July 1846 and again on May 22, 1847, summoning her to appear before the November 1847 term. Two years after Pettit filed his lawsuit, Julia Thurmond was still oddly unavailable.

So where were Henry and Julia Trammell? Had they left Arkansas with the slaves that were in dispute? It appears not, or at least not yet. Not long after his brother, Phillip's, death, Henry sold Phillip's nearby land and moved out of Lafayette County.[20] A month before the last summons, Henry gave G. D. Royston his power of attorney in Lafayette County, Arkansas, in September 1847.[21] Records indicate that Henry lived in southeastern Arkansas in Ashley County, near the Mississippi river, through at least 1850. His holdings there included thirty slaves and other assets.[22] Henry and Julia had too much property and land to be considered fugitives from justice. Why nearby Chicot County authorities could not locate Henry and Julia remains an unanswered question.

The Tavern Keeper, Nathaniel

Although Henry Trammell was feeling pressured by legal issues in Arkansas, it was Nicholas Trammell's oldest son, Nathaniel, who was the first member of the family to leave Arkansas and return to Texas. Nat's first wife, Leona (or Lena) Ward, died in Arkansas following a difficult childbirth in 1838. Nathaniel later married a woman about twenty years younger named Elizabeth (or Eliza) Lindsey.[23] Nathaniel and Eliza's first child was born in Arkansas in 1847, and a second child was born in Texas in 1849, so it is likely that Nathaniel had moved to Texas by 1848. Two female slaves, age forty and twenty-eight, and two young slave boys, age ten and five, were listed as part of Nathaniel's property. One of those boys, a ten-year-old named Frank, became a point of contention in a series of civil lawsuits that were typical for the Trammells after they moved to Texas.[24]

By the time of the federal census in 1850, Nat was settled in Fayette County near La Grange, Texas. The reasons the Trammells chose this part of Texas for resettlement is uncertain, but there are some likely reasons based on the family's history. La Grange was situated on the La Bahia Road, connecting with the El Camino Real between Nacogdoches and San Antonio. When Nicholas Trammell was actively trading, he likely traveled that route and perhaps found the area a good place for raising cattle and engaging in the family businesses of gambling and horse racing. Nearby racetracks at Bastrop, Goliad, Brazoria, Washington, Columbus, Crockett, and all along the Gulf coast also provided venues for running horses and betting on races.[25]

On August 24, 1852, Nat signed a one-year lease with Alexander Boyle to operate the Eagle Hotel in La Grange.[26] Nathaniel, his two daughters, and their husbands, George Washington Sawyer and Edmond Sawyer, all lived together on Nathaniel's ranch about eight miles outside of La Grange, on land Nat purchased in May 1852. Located at the corner of West Colorado and North Washington streets, the Eagle Hotel not only included room and board facilities but also a tavern, out buildings, and a livery stable, providing similar diversions to those the Trammells served up back in Arkansas—gambling at faro and an overnight stay with liquor available to grease the wallets of the gamblers.[27]

Nathaniel's first mention in the Texas courts was on November 1, 1852, when W. B. Branch dismissed a suit against him.[28] Over the next two years, Nathaniel Trammell appeared in numerous court cases in Fayette County, not only demonstrating that the Trammell family stripes had not changed but also that Nathaniel inherited his father's tenacity by failing to accept the initial judgment of any court and pursuing every appeal possible, even to the Supreme Court of Texas.

After opening the Eagle hotel in September 1852, Nathaniel Trammell promptly got back to the familiar family business of operating a tavern. One of the earliest mentions of Nathaniel in the courts describes his attempt to recover payment from a transient guest.[29] George W. Miles was a resident of the Eagle Hotel, where he boarded, lodged, and stabled his horse—or horses, since that is part of the dispute. When Miles did not pay his bill, Nat got a judgment against him from an acting justice of the peace in early 1853 in the amount of forty dollars. Miles still could not pay, so Constable James S. Patton of Fayette County sold Miles's sorrel gelding at auction on February 28, 1853. Miles filed suit against Trammell, claiming that the horse was unfairly acquired. The case hinged on whether the constable auctioned Miles's only horse or whether he had another. On March 20, 1853, Miles demanded that Nathaniel Trammell return the horse, but he refused. Miles now wanted $250 for the horse and $50 for damages from Trammell. The court record ends in the fall term of 1853, when Nat asked for a continuance in order to locate a witness. If the Trammell ways in court were continuing in Fayette County, Mr. Miles was in for a long case.

A more complex case involving Nathaniel Trammell began on April 22, 1853, in the Fayette County District Court. Judge John Hancock of Texas' second judicial district heard a plea from Alexander Boyle, the owner of the Eagle Hotel. Nat agreed to pay Boyle $600 per year in rent: $300 after six months and another $300 at the end of a year, terms that were "signed and sealed" and presented to a third party. By March 1, 1853, Nat was obligated to pay the first $300, but Boyle said Trammell had so far refused to pay. Now Boyle wanted his money with interest. The suit seemed simple: did Trammell pay, or did he not pay? But disputes with a Trammell were rarely uncomplicated.

Nathaniel Trammell's response, delivered by his attorney to the fall term of the court in 1853, demonstrated that he had received an excellent education from his father in the ways of legal challenges. He wanted the court to believe he had been wronged and injured. Nat addressed the facts in a limited manner, but his attorney went on relentlessly with four pages of utter indignation about the circumstances of the disagreement.

Nathaniel said he was the "keeper of a public house of entertainment in the town of LaGrange" whose reputation had been injured by Boyle's methods for gaining payment. On April 26, 1853, the court assigned Boyle a writ of attachment and the sheriff forcibly took possession of Trammell's young slave, Frank, one of Trammell's "principle waiters." Frank was twelve years old and apparently vital to the operation of the tavern. On the day the sheriff arrived to claim Boyle's property, Nathaniel was not at the tavern. Trammell's indignation was that no sincere attempt was made to contact him before taking his slave and, further, that Boyle accused Nathaniel of avoiding the court's judgment. Nathaniel Trammell aggressively denied that he "secreted or attempted to secrete himself for the purpose of evading the services of the ordinary process of law."

Trammell was not at his tavern because, he explained, his presence was required at his ranch eight miles out of LaGrange. Nat employed George Washington Sawyer, his son-in-law, to run the ranch. When Sawyer became ill and unable to manage the operation in the days before the sheriff knocked on the tavern door, Nat explained to the court that he "was compelled to give a large portion of his personal attention to the details and management of said farm, particularly on the 23rd, 25th, and 26th days of April 1853." Nat claimed that everyone at the tavern knew where he was and he could easily have been found. In particular, William G. Thomas, the clerk of the tavern, could have directed the sheriff to Trammell's farm but apparently did not offer up Trammell's whereabouts.

Trammell claimed Boyle was motivated by "hatred and malice" to "vex and harass" him. In his entire rebuttal, Nathaniel glosses over whether he indeed owed the money and goes quickly back to how he had been wronged. Nathaniel's counterclaim states that Boyle "injured him not only by said unlawful seizing but in his reputation and credit as a tavern keeper." Instead of owing Boyle $300, Trammell now wanted $500 for damages from Boyle and $1 per day rental for Boyle's possession of the slave, Frank. Trammell temporarily won back possession of the slave in this matter, but Boyle appealed and prevailed after a rehearing. Frank was again awarded to Boyle in December 1853.

Nathaniel Trammell made a final appeal to the Texas Supreme Court on January 14, 1854. This time the case was made on a different legal challenge that had nothing to do with Trammell's nonpayment of rent to Boyle. The testimony focused on a more complex Trammell family dispute over who actually owned Frank and his mother, Charlotte. Nathaniel Trammell came into possession of Charlotte before Frank was born. Nat's first wife, Leona (or Lena), died in 1838, and her father, Britton Ward, died in 1842. When his deceased wife's share of his estate was divided by lot among her siblings, Nat drew lot number three, which included possession of Charlotte. Testimony focused on whether or not Boyle could legally acquire Frank. The argument was based on whether Nathaniel Trammell actually owned him or if possession should have gone to his two daughters by his wife, Leona.

Thomas W. Harper lived a half mile from Nathaniel and knew Frank well. Harper testified that he heard G. W. and Edmond Sawyer, the husbands of Nathaniel's two daughters, say "if they owned Frank they would not give him for any other negro they knew of." A man named C. B. King testified for the court in Fayette County that Nat had offered to sell Frank to him eighteen months to two years before the suit began. There never seemed to be any question that the family behaved as if Nathaniel owned Frank, but Trammell and the Sawyers claimed Frank was the property of their wives and therefore not Boyle's to claim in payment of Nathaniel's debt.

The Supreme Court of Texas finally decided the case in 1858, four years after it was filed.[30] Nathaniel Trammell's attorneys convinced the Supreme Court that the appeals court made a mistake in granting Boyle possession of Frank. They remanded the case back for a retrial, but there is no record of the final decision.

May 1853 was a particularly busy month in the courts for Nathaniel Trammell and his son-in-law, G. W. Sawyer. On May 6, William M. Rice called in his debts against a number of people, including Nathaniel for $236.94 and Sawyer for $193.86.[31] The following day, Nathaniel Trammell was indicted for betting at faro. He pled guilty on May 11 and was fined only ten dollars. Sawyer pled guilty to keeping a gaming house and was also fined ten dollars. Evidently, there was a move afoot to crack down on gaming that was becoming too widespread. One man was charged with playing cards in a grocery.

The Good Brother, Henry

Henry Trammell appears to have once been the trusted brother. At least it would seem so when his brother Phillip appointed him executor of his estate and guardian of his son, also named Nicholas, back in 1844. Perhaps Phillip's 64-year-old father urged him to appoint Henry to care for his affairs when it became evident that Phillip would die within days. Henry's intentions may have been good at the time, but his actions soon resulted in a series of legal challenges to his handling of the estate and serious accusations against him by his family.

Although Phillip died in February 1844, Henry did not appear in probate court until November 1846. The accounting of Phillip's holdings indicated his relative wealth, with three hundred acres of rich farmland and about thirty slaves. When Henry appeared in probate court, he valued the estate at $36,331.94 at the time of Phillip's death. In the two years after Phillip's death, Henry had paid $3,668.36 of his brother's debts. Henry also paid himself a five percent commission as executor of the estate—over $1,500.[32] In addition, Henry reported that seven of Phillip's slaves had died since Phillip's death. The value of the reportedly deceased slaves was stated at $1,950. However, further evidence appeared that indicates Henry's statement about their deaths was false.

In November 1847, Henry received the approval of the probate court to sell some property to pay an additional $3,000 in debts. The court order permitting Henry to sell the remaining real estate and personal property was not issued until February 1848, four years after Phillip's death. Shortly after that order, Henry Trammell sold the land and other property to Richard F. Sullivan on March 22, 1848. Within a year Sullivan auctioned off the old Trammell place.[33] His newspaper advertisement made the pitch and provided some detail about Phillip's holdings: "The subscriber, will on the 10th day of February 1849 at his plantation, sell to the highest bidder on a credit of twelve months his plantation lying in Lafayette county between

Lost Prairie and Elam's Prairie known as Trammell's old place. There are about 300 acres in a high state of cultivation."34

On the same day and in the same newspaper notice, Henry Trammell indicated his plans to sell off the remaining slaves from Phillip's estate: "I will, as Executor of the will of P. Trammell, deceased, at the same time and place, sell to the highest bidder on a credit of twelve months, all the slaves, about 30 in number, consisting of men, women and children, house servants, and field hands, belonging to the estate of Phillip Trammell, deceased, or so many of them as will discharge all the debts of the deceased – the purchaser giving bond and approved security bearing interest at the rate of eight per cent, per annum, from the day of sale. Creditors of said estate are solicited to attend. January 24, 1849."35

Even though the probate court had given its approval to sell the slaves in February 1848, Henry had not taken any action until the time he advertised them for auction a year later. In the five years since Phillip's death, Henry had the benefit of the services of thirty slaves. In addition to the inevitable questions about the disposition of the property, the motivation for Henry's decisions with regard to young Nicholas became suspect as well. Phillip's will made his wishes very clear regarding his son: "The will of Phillip Trammell required that "the farm and negroes should be kept together," and the proceeds, after paying for the education and defraying the necessary expenses of the minor, should be invested in other property for him until he should arrive at twenty-one; and it was expressly directed by the testator that he should be kept until his majority "at one of the best colleges in the United States."36

These were the wishes of a dying man the day before he died. Phillip's death left his ten-year-old son, who was already without a mother, fatherless as well. Phillip was a wealthy landowner with slaves who had no doubt been with him since his son Nicholas's birth. House servants had cared for his family and kept his plantation in a "state of high cultivation." Now Henry was selling them off, and his use of the proceeds was in question.

When the outcome of the auction was discovered by the court, Henry's position as executor came under intense scrutiny. The slaves were reportedly auctioned separately rather than kept together as a family unit as Phillip specified in his will. Court records indicate the slaves were not sold but were instead hired out. Perhaps that is what Henry had already done in the five years since Phillip's death. Having slaves produce income for the benefit of an estate was a common and acceptable practice. The court that examined Henry's actions as executor noted that the revenue recorded for the estate indicated that they were rented rather than sold.[37] Henry confirmed as much shortly after the auction, when he requested an order from the probate court to sell the slaves and pay off debts that had somehow continued to mount long after he assumed responsibility for Phillip's estate. The probate judge grew tired of Henry's delays and continued failure to care for the estate and young Nicholas, as required by Phillip's will. The court refused Henry's request to sell the slaves and instead ordered that they continue to be hired out until the next term of the court.

It was not only the judge who was losing his patience. Thomas Hubbard was a former prosecutor, a prominent attorney in southwestern Arkansas, and later a circuit court judge.[38] At the November 1849 court term, Hubbard appeared in court to request that Henry Trammell be removed as executor of the estate. Although the record does not indicate whether Hubbard was acting on his own or on behalf of Nicholas Trammell, it seems apparent that his position supported Nicholas Trammell's disagreement with Henry's use of the estate for his own gain. Part of Hubbard's claim was based on a legal technicality—the fact that Henry was administering the estate without giving bond as the law required. Hubbard's complaint was void because that odd exception was actually allowed by Phillip's will. The more serious legal claim Hubbard made was that Henry Trammell was wasting and mismanaging the estate. Apparently, Hubbard's appearance in court was unexpected, and Henry Trammell was caught off guard. The court gave him until February 1850 to prepare his defense of the accusations.

At the second hearing, the probate judge found Hubbard's accusations to be correct. Henry Trammell was removed as executor of Phillip Trammell's estate. In an understated comment on the state of affairs surrounding this matter, the highest court in the State of Texas subtly observed that "at this point the facts cease to be clear."

A few months later, when the Arkansas federal census was conducted in November 1850, Henry was still in southeastern Arkansas, raising his three daughters; his nephew, Nicholas, Phillip's only child; and the three sons, Jarrett, Henry, and Robert, of his brother Robert, who died in January 1849.[39] Although he could not legally take any action with the estate, it did not mean Henry was not still responsible for Phillip's young son, Nicholas. Nor did it mean, at least in his mind, that he could not benefit from some of the assets—namely the slaves. In 1852, soon after being removed as executor, Henry Trammell and his extended family, as well as all of Phillip's remaining slaves, moved to Texas. Only the seventeen-year-old boy for whom he was appointed guardian, young Nicholas Trammell, stayed behind.

Henry moved to Fayette County, Texas, at about the same time Nat began operating the Eagle Hotel, and Henry must have joined him in the gaming operations. They were indicted together for related gambling offenses. On May 7, 1853, Henry was indicted for betting at faro, the same day and the same charge as Nathaniel. Nathaniel pled guilty four days later and was fined only ten dollars, but Henry delayed his response. Perhaps the unclaimed letter for him in the Washington, Texas, post office in August 1853 was the summons he was ignoring.[40] When Henry finally pled guilty to his faro charge on November 4, 1853, he was fined twenty-five dollars for his transgressions. The following day, Henry was assessed another twenty-five dollar fine after he pled guilty to the charge of playing cards.

Repeatedly operating a card game at the same physical location had its risks and could be easily discovered, so Henry augmented his gaming interests by engaging in the other Trammell family pursuit of horseracing. When Henry was accused along with Mote's son Harrison Askey of taking ninety-five head of cattle in December 1854, the men testified that they won the cattle in a horserace and sold them to a man named Creed Taylor.[41] On April 5, 1856, another dispute over a victim's payment of debt in horses made it to the courts.

Henry had bigger legal issues than the minimal fines for his participation in gaming operations. It was not Henry's departure from Arkansas that created a fresh set of legal problems within his family; it was young Nicholas's subsequent move to Texas with a new bride.

1853, April—Sadness and Disputes

The man for whom Trammell's Trace is named had been reported dead once. Nine years later, Nicholas Trammell read his name in an obituary for a second time. Back in 1844, a Houston paper confused the murder of a John Trammell of Arkansas with Nicholas Trammell. When a second newspaper account of the death of a Nicholas Trammell appeared in Texas in 1853, once again it was not about Nicholas Trammell the smuggler and trader from Kentucky and Arkansas and patriarch of the trading Trammell family.

Nicholas Trammell's grandson of the same name, Phillip's son, died of pneumonia on April 6, 1853, only two months after coming to Texas with his young wife.[42] Nicholas had married Amanda Holloway (or Halloway) back in Arkansas and moved to Fayette County, Texas, in February 1853. Nicholas died unexpectedly in the early spring, only eighteen years of age, perhaps nursing an illness acquired during his winter migration to Texas. Since he was still a minor, or "infant" in the legal language of the day, young Nicholas was legally still in the guardianship of his Uncle Henry Trammell. As the sole heir of his father's estate, Nicholas's lack of a will or children from his marriage raised legal disputes among the Trammell family about the disposition of the remaining slaves in Henry's possession. In January 1855, the senior Nicholas Trammell, his children, and other members of his family sued Henry and made serious allegations about his handling of Phillip's estate. He asserted that Henry Trammell "illegally and fraudulently wasted and converted to his own use a large portion of the estate aforesaid, and fled from the aforesaid state of Arkansas to the state of Texas, bringing with him a large number of the slaves belonging to the estate, viz.: the number of some twenty-five or thirty, and with the view of converting the same to his own use, and cheating and defrauding your petitioners out of their just rights, has, as your petitioners are informed and believe, sold and run out of this state, or caused to be done by other persons, the greater portion of the said slaves so brought with him to Texas."[43]

Between 1852, when Henry arrived in Texas, and 1855, when the lawsuit began, Nicholas Trammell accused his son of selling off twenty or more of Phillip's former slaves for his own gain. All but five were gone. The family's case against Henry was that he should have his own percentage of the value of the slaves, as one of Phillip's siblings, but should not have had sole possession and control. They argued that since Phillip's son, Nicholas, had no children or will, the remaining assets fall to his family members—not to his widow, Amanda, and certainly not solely to Henry.

When the suit commenced in 1855, Amanda Trammell appeared in court and joined in the defense with Henry Trammell. She was newly remarried in 1854 to Jarrett Trammell, one of Robert Trammell's sons who also lived with Henry. Amanda Halloway Trammell married her dead husband's cousin, thereby keeping the Trammell name.

The case that started in Gonzales County, Texas, in 1855 was not resolved until it reached the Texas Supreme Court in 1857.[44] The court upheld the decision of the lower court that Henry did indeed owe his family money. Beyond the terms of the settlement, the court was harsh in its judgment of Henry Trammell, stating that his whole involvement in the estate of his brother was fraudulent and shameful: "It is clear that his removal to Texas was a fraud, for he refused to obey the mandates of the court, and squandered property that did not belong to him. Nor can it be argued that he was the testamentary guardian of the boy. He should blush to call to his aid the will of the deceased brother, which authorized him without bond, to take charge of a large estate for an only child, and keep that child at college till he arrived at the age of 21 years. Instead of executing the will faithfully, he attempted to rob that child of his estate."[45]

There was division within the Trammell family. In his father's eyes, Henry had committed the ultimate disloyalty—treating family like the people they had taken advantage of in their trade and gambling operations. His whole life, Nicholas Trammell readily sought remedy in the courts when he believed someone was trying to gain an advantage—even if it was his own son.

With the court battle over the disposition of the slaves looming, Henry and his wife, Julia, anticipated the outcome and protected their assets. On May 11, 1854, Julia purchased 460 acres from Harrison Askey, Mote's son, in the J. K. Castleman headright on Sandies Creek, twenty miles southwest of Gonzales. When they made the purchase in her name, they were probably trying to separate their assets since they both had their own legal issues over the ownership of slaves; she with her Thurmond side, and Henry with Phillip's estate. Henry Trammell and Harrison Askey were landowners west of Gonzales. Nathaniel, and soon Nicholas Trammell, settled east of Gonzales. The family was together again in the same part of the country, but the divide between them grew wider after they moved to Texas.

1847-1854—Nick's Move to Texas

In the months and years preceding the patriarch Nicholas Trammell's relocation to Texas, business transactions hinted toward the move. Land transactions and the sale of other property document his disposal of anything that would make the trip more difficult. A man to whom Trammell owed money noticed the activity and began to worry about collecting his debt. A short article in the *Arkansas Gazette* on July 28, 1847, cautioned readers not to accept a note for one hundred dollars from Trammell, who owed the money to Thomas Woodward.[46] Woodward was a prominent citizen of Camden and had given the note to Jabez Smith for collection. Smith was described as a quiet, mild gentleman with a reserved style and one of the best judges of human nature around. Yet when stirred, he was a "lion who would forego demeanor for purpose."[47] Descriptions of Woodward placed him in a category of frontier businessmen who came into the region on the heels of men like Nicholas Trammell. The cultural differences between

Trammell and Woodward may have been as much a part of the dispute as the debt. Woodward was described as "a borderman of the west; not one of those who plunge into the forest and make their homes in and from it, but one of those who seek the outskirts of civilization, treading upon the heels of the former, partaking of their characteristics to some extent, and profiting by their hard and often blood-bought earnings."[48] Men like Woodward took advantage of men like Trammell, who had earlier taken advantage of settlers and Indians. If Woodward had "tread upon the heels" of Nicholas Trammell, having the rambunctious Smith seek collection of his debt from Trammell may have been a wise decision.

Tax records between 1847 and 1849 from Hempstead County, Arkansas, document Nicholas Trammell's ownership of 240 acres of land, livestock, and five slaves. The total value of his property was $3,850; $2,000 of that was the value of the slaves. Trammell sold three slaves in 1849 to Abraham and David Block. Becca, age twenty-one, and her three-year-old daughter and one-year-old son, stayed behind in Arkansas.[49] Between 1849 and 1850, Trammell's fortunes either increased or he was not completely forthcoming with the Hempstead County tax assessor. The federal census for 1850 showed Nicholas Trammell with eleven slaves between the ages of two and sixty.[50] In September 1849, one of Trammell's slaves ran away, perhaps not wanting to make a move to Texas: "Ran away from the subscriber, living in Hempstead county, Arkansas, in the month of September last, a negro man named Edmund. He is about 25 years old—a bright mulatto—has straight hair, and is about five feet ten inches in height. He reads well and speaks quiet. Said boy was raised in this state. The last that was heard from him he was in Dallas county in this state. He is probably endeavoring to get to Missouri or Illinois. I will give fifty dollars reward for his apprehension and confinement in some jail, so that I can get him; or one hundred dollars if delivered to me."[51]

Why Trammell waited almost a year after Edmund ran away to advertise his departure is unclear. Both Edmund's disappearance as well as Trammell's attempt to gather him back may have been a result of his preparations to move back to Texas.

Nicholas Trammell was enumerated as a fifty-nine-year-old farmer by the 1850 Hempstead County, Arkansas, census, both slight deviations from the facts since he still operated the tavern and was sixty-nine years old. His wife, Mary Sadberry Trammell, was only thirty-eight years old at the time, thirty-one years younger than her husband. Perhaps that difference led to Trammell's report of a younger age. The census noted the two children of their marriage. Nicholas, age five, and Sarah R., age four, were both born in Arkansas. A woman named Mary Baily, age thirty, also lived with the Trammells, but nothing is known of her presence there or her connection to the family.

Sometime in 1850, Nicholas Trammell celebrated his seventieth birthday. Nathaniel had already moved to Texas, and Henry's battles with the probate court put the thought of his own removal to Texas at the forefront. Two of Nicholas's sons were dead, as was his first wife. His much younger second wife, Mary, may have been the one to make the decision to leave Arkansas behind. It is difficult to imagine what would pull Nicholas Trammell away from his longtime home in Hempstead County other than the desire to be near his extended family or the fears of being alone in his old age. Trammell lived at the crossroads in Hempstead County along the Washington-Camden road for over twenty years, longer than anywhere in his entire life. Precisely when Nicholas Trammell, his wife, and their two young children made the trip down Trammel's Trace to Nacogdoches and along the El Camino Real to the area around La Grange and Gonzales County is not certain, but records indicate it was probably in 1853 or early 1854.

Trammell's turn toward Texas culminated with his sale of the land he had occupied in the latest years of his life. Trammell sold the eighty acres he acquired from Isom Parmer to James M. Hancock in the spring of 1852.[52] Trammell was also gathering up stock, not only his but also unclaimed stock roaming the area. He filed two estray bonds in 1852, a process by which he could claim loose stock one year from the filing.[53] In what may have been one of Trammell's final transactions before leaving Arkansas, he sold his hogs to Hancock in April 1853 along with his registered livestock brand.[54] In exchange for the hogs, which would have been very difficult to move across hundreds of miles of trail to Texas, Trammell received twenty-five bushels of corn. Trammell was now ready to make the move to Texas to be near his remaining family.

Nicholas Trammell had moved property and family to Texas once before, herding his cattle and horses down Trammel's Trace from Pecan Point. That effort was to try to settle in Austin's colony along with others from the Red River. When Austin would not accept him, Trammell found an alternative at the Trinity ferry crossing on the El Camino Real. Before Trammell could turn that key intersection into a haven for traders and gamblers, he was chased out of Texas at gunpoint. Back in Arkansas he found a way to make a living from a stationary point rather than by moving goods down the trail named for him. When Nicholas Trammell took one last look back across wagons loaded with his possessions to the cabin and tavern where he had spent over twenty years, raised his family, and lived with two wives, he knew it would be for the last time. His wife had Sedberry relatives who lived nearby in Hempstead County, and they also made the move to Gonzales and Fayette Counties in Texas. There is no way to know why all the Trammells and their extended kin moved together to Texas. Just as they had with every migration, when the Trammells and their kin moved, they continued living in close proximity. On December 9, 1854, Nicholas Trammell bought 295 acres, six miles east of Gonzales along Denton Creek.[55] It was there that his life soon came to an end.

1856—Nick's Death

After two false reports of Nicholas Trammell's death, a third account finally focused on the man whose trail had delivered so many into the new frontier.

Trammell fell seriously ill with symptoms of pneumonia, but it may have simply been his many years of living that finally caught up with him. Seventy-six years was a long life in the 1850s. The average life expectancy was much shorter, and for people living on the edge of the law and the fringes of civilization, life could be abbreviated for all sorts of reasons. The loss of his first wife, two sons, and his half brother were testament to that.

A Dr. Nicholson, the physician who had been treating him since around July 1855, paid a final visit to the Trammell home on April 12, 1856. Medical practice in the 1850s for patients with pneumonia, or what was then called "bilious pleurisy," called for the use of bloodletting. Bleeding was initiated by the use of leeches, lancets, or, in Trammell's case, the scarifying and cupping procedure.[56] In this method, an instrument called a scarifactor was used to cause bleeding, usually on the arm. A metal device about five inches square held multiple spring-loaded blades which quickly sliced lacerations through the skin when the spring was released. The effect was that of multiple slices with a sharp edged razor. The openings were then covered by a small metal cup that was placed over each wound and heated by flame from a cupping lamp. The vacuum created by the process filled each cup with blood and caused painful blisters. Medical practitioners believed that removing significant volumes of blood through repeated bloodletting, though painful, cleared the system of the bilious invaders.[57]

As Trammell lay there on his bed for months, suffering both from the symptoms of old age and illness as well as the severe treatment administered in response, one cannot help but wonder how he considered his life. Although often credited with violence, there are no direct indications that he carried out such acts. His efforts at clearing the roads on which he and others traveled for commerce and immigration were made for the purpose of trade or for an opportunity for a new and better life. His many lawsuits demonstrated a tenacious pride. Racing and gambling pursuits demonstrated a competitive and determined character, although one not opposed to relieving another man of his money within the norms of those vices. Given his lengthy decline and the suffering he no doubt endured, we might assume that Nicholas Trammell Jr. was indeed prepared for death, perhaps even welcoming it.

The doctor's efforts were ineffective. Nicholas Trammell died in April 1856, in Gonzales County, Texas, far from Tennessee and Kentucky and from the woods of Arkansas and Texas where he made his name.

Oddly, Nicholas Trammell died without a will, but his wife Mary filed an inventory on June 30. His land on Denton Creek was 247.5 acres. Four slaves were worth $3,000, and four horses were valued at $230. Sixteen cows, two yoke of oxen and a hundred hogs were included in his livestock. Two wagons, a shotgun, a rifle, and miscellaneous tools and kitchen articles rounded out the list. The services of the good Dr. Nicholson were also listed with Nick's debts. Seventy-eight dollars were owed for medicines and treatment. Although a value could not be placed on it, he also left behind an old road from Arkansas to Texas that old survey maps still remind us was a road between the past and the future of an entire region and its people.

10

1856–1880
The Patriarch Has Passed

> *"They shut the road through the woods*
> *Seventy years ago.*
> *Weather and rain have undone it again,*
> *And now you would never know*
> *There was once a road through the woods . . . ."*
>
> —RUDYARD KIPLING[1]

The Transformation from Life to Legend

Over one hundred fifty years after Nicholas Trammell's death, there is still much to correct about historical judgments of the man. Nicholas Trammell has been the subject of myth, legend, and misinformation, a process that started while he was still alive. Trammell's business ventures and personality always stimulated talk. The Cherokee accused him of inciting them to murder and burn the houses of his Anglo neighbors along the White River. Authorities in Nacogdoches chased him out of Texas at gunpoint. Folks around his notable Arkansas tavern on the Washington-Camden road talked about his suspicious activities at horseraces and taverns.

There was talk about where he went, what he did, and where his money came from, but it was always talk. Other than his temporary incarceration in Kentucky, there were no criminal convictions, no newspaper accounts warning of criminality, nothing to indicate that he was the villainous murderer described in the legends that developed a hundred years later. Nothing but debts and lawsuits over financial or trade issues appeared in court documents, but there were enough of those to extend even beyond his lifetime.

Much of the legend of Nicholas Trammell can be traced to the repetition of two accounts that came long after his death. In one, an interview with a former slave wove a tale of kidnapping and murder high on drama but difficult to assign to fact. In the other, there were rumors of Trammell as a prescient overseer who silently ruled others for his own gain. When Trammell left Arkansas and moved back to Texas in 1853, he left open the doors for future storytellers to embellish his role. Those tales, exaggerated by their distance from anyone who may have once known the facts of his life, led to the growing myth and legend surrounding the notorious Nicholas Trammell.

One of the legends was a result of a program that was part of Roosevelt's New Deal. In 1936, an interviewer with the Works Progress Administration (WPA) interviewed Robert Samuels, a former slave known as Uncle Bob, and chronicled a tale that was part fact and part fiction. Samuels served as a county clerk in Hempstead County during Reconstruction. He was quite old and blind at the time of the interview but spoke clearly. Uncle Bob told a story about how his mother and his grandmother were of mixed Spanish blood and came to Texas by way of Gaines Ferry across the Sabine River, east of Nacogdoches. Uncle Bob claimed that his grandfather was killed by Nicholas Trammell and John Murrell, an outlaw of some notoriety, and his grandmother forced to marry against her will. Samuels's story, though frighteningly compelling, suffered from difficulties of memory and fact. However, there are facts that appear to support that both his mother and he were once part of the Trammell family's slaveholdings.[2]

The second story suggests more redeeming traits for Trammell but also provides an accusatory judgment of the defects of his personality. Sam Williams was a newspaperman in the Arkansas Territory and an agent for the *Arkansas Gazette* after Trammell left Arkansas. Williams was a contemporary of key figures around Hempstead County who could have provided him firsthand or secondhand tales of Nicholas Trammell's life in Arkansas. One of Williams's stories, written around 1886, involved Trammell and Joshua Morrison.[3] When this reported event actually occurred is unclear, but Joshua Morrison lived at Chickaninny Prairie in 1827, where he was elected the first sheriff of Lafayette County. In that capacity, Morrison arrested Trammell in a civil matter regarding a debt owed to John M. Bradley, so Morrison and Trammell not only lived in the same region, but Trammell would have been familiar with Morrison's badge as sheriff.

Williams's story told how Trammell helped Morrison secure a land title by lending him a faster horse to get to Little Rock in advance of another potential claimant. The tale is complete with dialogue and details only the two men present would have known. It leaves the impression that Williams may have begun with a limited set of facts and then embellished them with his writer's pen. Some of what Williams wrote about Trammell was the estimation of others, but his words also contained Williams's opinions about what he had heard.

> He [Trammell] was an odd, exclusive, secretive sort of fellow; mingled but little with his neighbors, and, when he did, was noted more for what he did not than what he did say; was rarely at home to any one, and as rarely called at other people's houses. His comings and goings were secret and mysterious, being frequently conducted under cover of darkness. He would quietly disappear, remain absent for days, neighbors knew not where, or upon what business or mission. Then as unexpectedly he would turn up again, and for a season pursue the even tenor of his way about home. He always appeared to be flush of money, without seemingly putting forth any great effort, either of muscle or brain power, for the procuring of it. This fact, together with the queer movements and absolute unsociability of the man, gave rise to a good deal of gossip and speculation concerning his character and doings.[4]

Williams acknowledges that there was substantial gossip about Trammell, and with good reason. Although the description would have suited any gambler and trader, it seemed to become certain proof that Trammell was surely up to no good. Part of the myth was that Trammell belonged to a gang of counterfeiters and horse thieves and that he had been a confederate of John A. Murrell in the days when that notorious land pirate was the terror of travelers in the southwest. There were reports of strangers coming and going from Trammell's house—rough and desperate looking fellows—altogether different in appearance and actions from reputable and law-abiding men.[5]

Williams frankly admitted that he never knew the extent of the truth in his reports. He was a newspaperman but writing stories by standards that allowed for embellishment and drama. Williams's father operated two taverns in Fulton, so the stories about Trammell that Williams heard there probably flowed with the aid of liquor. Williams simply accepted that everyone in the country had heard the tales repeatedly, so they must have some basis in truth. In Williams's view, if "old Nick never took the pains to deny or correct them," they must be true.[6]

Beyond the tales of Trammell's movements, Williams offered a personal assessment of Trammell's physical stature and personal foibles: "Trammell was a small framed, non-combative looking sort of a fellow. In general appearance he was the very antithesis of the typical Western desperado. Personal courage, it is my opinion, was totally absent in him. I never heard of his being engaged in a fight or of offering to resent an insult. His legs, though quite small, usually exerted a controlling influence upon his actions when such emergencies as these confronted him. He no doubt possessed the intelligence and shrewdness to successfully plan depredations upon life and property, but the physical courage to execute them was lacking by him, I am sure."[7]

Williams's depiction of Nicholas Trammell as a man who avoided physical fights certainly matches accounts of his conflicts. Whether he was challenged in court or at the end of a gun, Nicholas Trammell always lived to fight another day and prevailed through the courts rather than with a vengeful attack.

Seventy-five years after his death, stories written with high drama created a sense of mystery around Trammell and his tavern. In language sounding like a promotion for a Hollywood thriller, writers talked about Trammell's inn as a place where people checked in but did not check out: "If old stories handed down by word of mouth are true—and one instinctively feels that they are—few travelers who 'put up' for the night at Nick Trammell's place ever lived to see the sight of a subsequent morning."[8]

It would be quite a stretch to think that the opportunistic Nicholas Trammell killed off the people from whom he made his living, much less that such murderous behavior could escape the law. His business relied on repeat customers and word of mouth. Bettors relieved of their money once would be welcomed back again and again. That the stories do not appear to take hold until thirty years or more after his exit from Arkansas does not add credibility to their dramatic detail. Nicholas Trammell was no doubt secretive and able to take advantage, but little in his history suggests that he was a horrific predator.

Descriptions of Trammell's cabin many years after its abandonment created a sensation of fear and foreboding at a place that remained a landmark along the route for years after Trammell left. In December 1862, the Hempstead tax collector spent two days at "the old Trammell Place, Carcuse township" to accept payment of property taxes for the coming year. The site of Trammell's bustling inn was by 1937 called an "eerie spot well envisioned for [murder]." The two old oaks that framed and shaded the cabin became "lonely sentinels" along the road where old wagon tracks were once still etched into the surface. What once was open farmland was by then cloaked by brambles, briars, and blackjacks in an "aura of melancholy, stillness, and gloom, enshroud[ed] with an air of everlasting mystery and remoteness." One story of the remains of his former tavern and home sounded like a tale meant to frighten away curious youth. It even included the presence of "blood marks—or what appeared to be blood marks—found on the walls of the two-story Trammell dwelling." A poem in a local newspaper appeared on Halloween Day in 1941, adding to the myth in celebration of the holiday reserved for such tall tales.

> The heartless keep had a special room
> And in it many travelers met their doom.
> Not satisfied with money, he took life;
> And slashed his victims with a gleaming knife.
> The bloody stains remain upon the wall,
> And a thick, uncanny silence covers all.[9]

It seems unlikely that Trammell would have avoided criminal prosecution if such wanton violence and disappearances were associated with his tavern. Trammell's behavior was guided by gaining a commercial advantage, not by murder or theft. Traders and tavern operators had little motivation to eliminate the very source of their prosperity. Particularly in the 1930s and 1940s, local lore wanted to make Trammell a scary, murderous killer. Trammell was a businessman, not a cold-blooded killer.[10]

The Family Goes on

When Nicholas Trammell died in Texas, he left behind sons Henry and Nathaniel, as well as Harrison Askey, the son of his half brother, Mote Askey. Trammell's daughter Nancy was in Texas as well, by then the wife of Green Baker. His two children of his second marriage, another Nicholas and Sarah, remained in Texas with their mother, Mary Sadberry Trammell. Mary had Sedberry family nearby in Gonzales.

Nicholas Trammell left behind a contentious and disagreeable family. His heirs and descendants continued their court battles with each other over slaves and property for well over twenty years. The Texas Supreme Court case styled *Trammell v. Trammell*, a dispute over ownership of a slave, continued in the courts for years. Henry Trammell's remaining siblings and his stepmother were pitted against him in a disagreement about his management of his brother Phillip's estate, primarily ownership of his slaves. Nicholas Trammell's second wife, Mary, was relentless in her pursuit of that claim. Henry finally lost the case, and a judgment was made against him not only with regard to ownership of Phillip's slaves but also for the gain he received from their rental and sale. Mary's attorney, Benjamin Shropshire, collected over $4,000 from Henry and others who posted bond. The case began while Nicholas Trammell was still alive, so following his death, Shropshire was uncertain to whom the money should be released. Mary Trammell finally had to sue Shropshire in order to collect the settlement.[11] Even though Henry and Julia Trammell bought land in 1858 that gave them considerably more property than anyone else in the family, by November 1859, when Henry was ordered to pay the settlement, he informed the court he had been "pursued to insolvency" by his own family.[12]

After Nicholas Trammell's death, his wife and the rest of his family stayed in Texas in the area around Gonzales and La Grange. In the 1860 census, Trammell's wife, Mary, and his son Nicholas, age seventeen, were still residents of Gonzales County, Texas. Ten years later, in 1870, they were in adjoining Fayette County living between Hallettsville Road and Buckner's Creek. Nathaniel Trammell, age sixty-three, and his wife Elizabeth Davidson Trammell, age fifty-three, remained in Gonzales County. A twenty-five-year-old woman also named Elizabeth Davidson and her young son, Alexander, age one, lived with them. By 1880, almost twenty-five years after Nicholas Trammell's death, his wife Mary, then seventy, was living with the Sedberry side of her family in Gonzales County. Henry and his family were there as well, including four young children under the age of eight.

Save, Oh Save, the Road

The expansion of roads into and across Texas continued unabated. The route that Nicholas Trammell created for undetected movement was but one of a network of roads winding between new settlements. The El Camino Real remained the most heavily used. A year after Trammell died, road conditions along the El Camino Real continued to worsen due to the increased immigration and travel: "The road [from Natchitoches to San Augustine] could hardly be called a road. It was only a way where people had passed along before. Each man had taken such a path as suited him, turning aside to avoide, on high ground, the sand; on low ground, the mud. We chose generally, the untrodden elastic pavement of pine leaves, at a little distance from the main track."[13]

In 1858, there were three roads between Camden and Washington, Arkansas, not just the one that passed by Trammell's tavern. One road passed Prairie de Ann (near Prescott), another through Poison Springs and Rocky Mound (near Trammell's place), and a lower road by Moscow and Prairie DeRoane. By 1859, for about ten cents a mile, ambitious seagoing travelers could enter Texas at Galveston or Houston and travel up the Trinity River by steamboat to the El Camino Real near Crockett. From there, four-horse stagecoaches went to Nacogdoches and from Nacogdoches to points all across east Texas and beyond. Up Trammel's Trace from the old Spanish outpost, the stage ran through Mt. Enterprise, Camden, and Marshall.[14] Travel was safer and more certain by that time as well. With the exception of the Alabama and Coushatta, virtually all of the Indians were gone from East Texas by 1859.[15] By 1860, only a few miles of roadway in all of Texas were graded, and only twenty miles were plank road.[16]

During the Civil War years, the old Trammel's Trace was still being used—its reputation spread by family who used it to enter Texas years before. In 1864, a Civil War regiment marched over some bad roads between the Red River at Fulton to the Camden ferry crossing of the Sabine River between December 3 and December 9. Confederate general Jo Shelby followed Trammel's Trace into Texas in 1865, leading one hundred troops all the way into Mexico when he refused to accept the victory of the North.

Immigration had no respect for the terrain. Rutted roads only created parallel routes, and fallen trees or high water led to more turnouts. When routes got so bad that they could not be used, there were enough people around with interests in making a new road that travel continued to key commercial locations. The reasons that Trammel's Trace came into being were the same reasons that new roads continued to develop across the region. It was trade and commerce that moved men now, not the exploration of new territory.

Even into the current era, sixteen years after the new millennium, roads follow other roads, and some of those roads follow along or near the route of old Trammel's Trace. Trammell could not have imagined the acres of concrete in today's modern roadways. He could not have foreseen how his early horse smuggling could have led to such a monstrosity of commerce carried on at speeds beyond his imagination. The scope and scale of today's highway transportation system would have been incomprehensible to a horse trader and gambler from Kentucky.

Nevertheless, Trammell the businessman would doubtless have recognized the commercial forces at work. He would have seen and understood the opportunity, just as he did over two hundred years ago.

Nicholas Trammell was not a visionary. He was just a horse smuggler who changed a part of history.

NOTES

Chapter 1

1. Helen P. Norvell. *King's Highway, the Great Strategic Military Highway of America, El Camino Real, the Old San Antonio Road*. (Austin: Firm Foundation Publishing House, 1945.)

2. Caddo Mounds State Historic Site is west of Alto, Texas, on the path of the El Camino Real. An excellent history of the Caddo is available online. See the bibliography for University of Texas, "Tejas: Life and Times of the Caddo." For a complete academic history of the Caddo, see Timothy K. Perttula, *The Caddo Nation*, 1992.

3. Jeffrey M. Williams. *GIS Aided Archaeological Research of El Camino Real De Los Tejas with Focus on the Landscape and River Crossings along El Camino Carretera*. (Nacogdoches: Master's thesis, Stephen F. Austin State University, 2007), 27.

4. See work by Herbert Eugene Bolton. *Athanase de Mezieres and the Louisiana-Texas Frontier, 1768–1780: Documents Pub. For the First Time, from the Original Spanish and French Manuscripts, Chiefly in the Archives of Mexico and Spain*. (Cleveland: Arthur H. Clark Co., 1914.)

5. Bolton, *Athanase de Mezieres*, 166.

6. Betje Klier Black. *Pavie in the Borderlands: The Journey of Theodore Pavie to Louisiana and Texas, 1829–1830, including Portions of His Souvenirs Atlantiques*. (Baton Rouge: Louisiana State University, 2000), 208. The phrase used here is from Pavie's essay on forest fire.

7. In 2004, a 550-mile section of the El Camino Real leading from Natchitoches, Louisiana, to the Rio Grande River was designated as a National Historic Trail named El Camino Real de los Tejas. In this book, "El Camino Real" refers to what is now the National Historic Trail named El Camino Real de los Tejas.

Chapter 2

1. R. L. Jones, "Folk Life in Early Texas: The Autobiography of Andrew Davis," *The Southwestern Historical Quarterly* 43 (July 1939–April 1940): 161.

2. Fray Jose Maria de Jesus Puelles' *Mapa Geographica de la Provincias Septentrionales de esta Nueva Espana of 1807* can be found in the Dolph Briscoe Center of

the University of Texas in Austin. For more discussion of this trace to the Spanish Bluff, see Dan Flores, *Southern Counterpart to Lewis and Clark: The Freeman and Custis Expedition of 1806.* (Norman: University of Oklahoma Press, 2002), 191–92.

3. Although Jonesborough and Pecan Point are two distinct points on early maps, it was Pecan Point that was the earlier settlement and most often referenced by traders and Indian agents. When mentioning the access points from there to Trammel's Trace, I use them both together to indicate a focal terminus to what came to be called Trammel's Trace. When referencing activity in each locale, they will be noted separately where the documents indicate.

4. Pris Weathers, "Arkansas Ties: Town of Fulton—1819." An ad in the *Arkansas Gazette* on December 25, 1819, reads: "The Town of Fulton is handsomely situated off the north-east bank of main red river, about two miles below the mouth of Little red River, and near the centre of the boundary of Arkansas Territory on Red River. It is the principal landing and place of deposit, for the country of Hampstead [sic], and is also the point at which the most direct road leading from Missouri territory, and many of the eastern states, to the extensive and fertile province of Texas, will cross said river." Her website, www.arkansasties.com, catalogs hundreds of articles from the *Arkansas Gazette*, a regional newspaper of the times.

5. James C. Martin and Robert Sidney Martin. *Maps of Texas and the Southwest, 1513–1900.* (Austin: Texas State Historical Association, 1999), 121. The White, Gallaher, and White 1828 "Mapa de los Estados Unidos de Méjico" (Martin, 137) clearly showed the path later called Trammel's Trace and included both northern branches to the Red River, one northwest to Pecan Point and the other leading northeast to Fulton, Arkansas.

6. Jones, *Autobiography*, 323.

7. Library of Congress. *Maps, American Memory.* http://memory.loc.gov/ammem/. The American Memory Project of the Library of Congress provides online access to several different maps of Texas, Arkansas, and Louisiana that show trails that became part of Trammel's Trace. By the 1830s, more direct routes between settlements replaced the trace as the primary route for routine.

8. William C. Davis. *A Way through the Wilderness: The Natchez Trace and the Civilization of the Southern Frontier.* (New York: Harper Collins, 1995), 14. Davis provides a description of conditions on the early Natchez Trace using a journal written by an Englishman named Francis Baily who left Natchez, Mississippi, in 1797 bound for Nashville. His journal provides some insight into the perils of travel during that time.

See also Josiah Gregg, *Commerce of the Prairies, or, The Journal of a Santa Fe Trader: During Eight Expeditions across the Great Western Prairies, and a Residence of Nearly Nine Years in Northern Mexico.* (New York: H. G. Langley, 1844.) This is an excellent resource on travel conditions in Texas during this time period.

See also Jack Jackson, ed. *Texas by Terán: The Diary Kept by General Manuel de Mier y Terán on his 1828 Inspection of Texas.* (Austin: University of Texas Press, 2000.)

9. Southwest Trail. http://www.southwesttrail.com. This route from the northeastern corner of Arkansas to the southwestern corner at Fulton and the Great Bend of the Red River was often referred to as the Old Military Road or the Na-

tional Road. In this book, it will be called the Southwest Trail, which is more presently used to name the collective group of trails and roads that roughly followed the edge of the Gulf Coastal Plain across Arkansas.

10. Larry Priest (ed.) and Kathryn Priest. *The Diary of Clinton Harrison Moore.* http://www.pcfa.org/genealogy/ClintonHarrisonMoore.html. The citation is for Moore's diary entry for March 18, 1839.

11. Where applicable, sources to support the following depictions of the terrain and difficulties of travel are cited. Del Weniger's landmark work, *The Explorers Texas: The Land and Waters*, provides many details for depictions of the route to follow. Since the route of Trammel's Trace has been generally established, information about the landscape can be described with relative accuracy. A mud hole is still a mud hole and a rattlesnake still fearsome. The author's wish is to convey some sense of the emotion of such a difficult journey, and he requests of the reader the liberty to explore that without direct documentation.

12. Daniel Davis says Trammell was the first one to clear this route in about 1821.

13. This route to the El Camino Real afforded the relative safety of remaining inside the United States until reaching the Sabine River crossing at Gaines' Ferry. Trammel's Trace reportedly had better water resources.

14. Jimmy Oliphant, ed. *Jeremy Hilliard: Gone to Texas.* (Marshall, Tex.: Self-published, 2000.) Publication housed at Harrison County Historical Museum, Marshall, Texas. See letter from Jeremy Hilliard to Mrs. Elizabeth Toole, "Camp on the Battlefield at Guilford, NC, six miles west of Greensboro," written on September 26, 1848.

15. Jones, *Folk Life*, 161. Andrew Davis called these packs "ciaxes."

16. James Dawson and Mary Eakin Dawson. *Trammel's Trace*. The Dawsons researched and mapped Trammel's Trace during the 1940s. A manuscript of their work, along with the definitive map of Trammel's Trace produced using survey information, is in the collection at the Southwest Arkansas Regional Archives in Washington, Arkansas. This reference is from their manuscript note 14, citing the *Arkansas Gazette*, Hempstead term of court, 1820.

17. "Vicinity of Fulton, Ark's," in Jeremy Francis Gilmer Papers, Gilmer Map Number 180, May 20, 1864. http://www.lib.unc.edu/mss/inv/g/Gilmer,Jeremy_Francis.html.

18. Accounts of the difficulty of water crossings can be found in various sources. A diary entry by P. V. Crawford on crossing the Snake River is dated August 10-11, 1851, accessed December 30, 2013, at http://tomlaidlaw.com/otkiosks/snake.html. See also George W. Riddle, *History of Early Days in Oregon.* (Riddle, Ore.: Riddle Enterprise, 1920), 14. Accounts of crossings from diaries by Mrs. Marcus Whitman, August 13, 1836, and Elizabeth Wood, August 21, 1851. http://www.traveltipsandtricks.com/rv_adventures.shtml. Accessed December 30, 2013.

19. George William Featherstonhaugh. *Excursion through the Slave States, from Washington on the Potomac, to the Frontier of Mexico; with Sketches of Popular Manners and Geological Notices.* (New York: Harper, 1844), 124. Featherstonhaugh gave this 1834 account of a crossing near the location of Dooley's Ferry, south of Fulton.

Even with the benefit of a ferry, his crossing into what the ferryman called "Spain" (it was Mexico at the time) was treacherous.

20. "Vicinty of Fulton, Ark's."

21. "A Visit to Texas in 1831, Being the Journal of a Traveller through those Parts Most Interesting to American Settlers, with Descriptions of Scenery, Habits, etc," *Houston Post Dispatch* (1929): 192.

22. *Encarnacion Prisoners: Comprising an Account of the March of the Kentucky Cavalry from Louisville to the Rio Grande, together with an Authentic History of the Captivity of the American Prisoners, including Incidents and Sketches of Men and Things on the Route and in Mexico.* (Louisville: Prentice and Weissinger, 1848), 8–13. See also William McClintock, "Journal of a Trip Through Texas and Northern Mexico in 1846-1847," *Southwestern Historical Quarterly* 34 (July 1930–April 1931): 20–37.

23. *Encarnacion Prisoners*, 11.

24. Harry Lee Williams. *History of Craighead County.* (Greenville, S.C.: Southern Historical Press, 1977), 149–56. The author provides a detailed and vivid description of various types of Indian trails.

25. When the boundary between the Republic of Texas and the United States was marked in 1840-41, eight-foot-tall dirt mounds were constructed every mile along the line north of the Sabine at Logansport, LA. Two roads crossed near there, Trammel's Trace and the road from Dooley's Ferry. United States Congress. *A Collection of Maps, Charts, Drawings, Surveys, etc, Published from Time to Time, by Order of the Two Houses of Congress (United States, Western Hemisphere, and the World).* (Washington: United States Congress, 1843.) http://www.loc.gov/resource/g3700m.gct00284/?sp=147.

26. Gregg, *Commerce*, 366.

27. This describes the route of Trammel's Trace just northeast of present-day Redwater, Texas.

28. Williams, *GIS Aided*, 169. The Lobanillo Cuts on the El Camino Real west of Nacogdoches retain the effects of heavy trail use and have seven parallel ruts.

29. *A Visit to Texas*, 205–207.

30. Featherstonhaugh, *Excursion*, 8–132.

31. Klier, *Pavie in the Borderlands*, 153.

32. Frederick Marryat, *The Travels and Adventures of Monsieur Violet in California, Sonora, and Western Texas.* (New York: George Routledge and Sons, 1843), 358. Marryat visited the United States in 1839.

33. Rex W. Strickland, "Moscoso's Journey through Texas," *Southwestern Historical Quarterly* 45 (1942): 109–37.

34. "An Act to Authorize the Post Master General to Establish a Post Route." *Laws of the Republic of Texas*, approved December 18, 1837, signed into law by Sam Houston.

35. As a result of investigations by the author and colleagues, this crossing was identified. The efforts of Bob Vernon of Bivins, Texas, led to a Texas state historical marker at this location on Highway 77 near Dalton Baptist Church.

36. This intersection became part of the Daniel Barecroft headright survey near the location of Old Unionville and present-day Naples in northwestern Cass County.

37. This section of the road had many iterations. It was first referenced in surveys as the "Mexican Trace," but part of it was the road the Spanish used to intercept the Freeman and Custis expedition in 1806. It also became part of the Jonesborough to Nacogdoches Road and was referred to in one land survey as Dayton's Road.

38. George Wilkins Kendall. *Narrative of the Texan Santa Fe Expedition Comprising a Tour through Texas and Capture of the Texans.* (London: Wiley & Putnam, 1844), 111.

39. Jones, *Folk Life*, 324.

40. Jones, *Folk Life*, 327.

41. The old Indian village was about a mile east of what is present-day Hughes Springs, near where Highways 11 and 49 intersect. Some years later, a dispute over the ownership of this land, which became the Joseph Burleson headright survey, was a landmark case in the Texas Supreme Court. See *Urquhart v. Burleson.* Supreme Court of Texas. 6 Tex. 502. 1851. *Reports of Cases Argued and Decided in the Supreme Court of the State of Texas during a Part of Galveston Term, 1851, and the Whole of Tyler Term, 1851, Vol. 6.* 251–57.

42. William B. Dewees. *Letters from an Early Settler of Texas.* (Louisville: Morton & Griswold, 1852), 22. Dewees states that he saw the Indians on Boggy Creek, but it is unclear which of the many streams named Boggy Creek that might have been.

43. Near the present town of Jefferson, Texas.

44. Klier, *Pavie*, 18.

45. Betje Black Klier. *Tales of the Sabine Borderlands: Early Louisiana and Texas Fiction by Théodore Pavie.* (College Station: Texas A&M University Press, 1998), 99. In those days, copal was an item of commerce, used for its sweet aroma.

46. This crossing would later be the site of Ramsdale's Ferry, where Harrison, Rusk, and Panola counties intersect at the Sabine River.

47. Gregg, *Commerce*, 360.

48. This is near Shiloh Baptist Church, just off Highway 315 northeast of Mt. Enterprise, Texas, not far from where the author's ancestors settled and the family still has land.

49. Samuel Botsford Buckley. *Geological and Agricultural Survey of Texas.* (Houston: A. C. Gray State Printer, 1874), 92.

50. W. T. Block. *East Texas Mill Towns and Ghost Towns, Vol. 1.* (Lufkin: Best of East Texas Publishers, 1994), 146. Quoting an article from the *Beaumont (TX) Journal*, November 13, 1904, Block reported a long-leaf pine with 283 tree rings.

51. The path of Trammel's Trace no longer appears on the old maps from the Texas General Land Office at a point just north of present-day Mount Enterprise. The reason for the sudden ending of its demarcation is not known. The location of the old trail from that point to its end at Nacogdoches about twenty-six miles farther south was unmapped on Texas headright surveys. However, roads across early Spanish land grants are the likely route, and a comparison with land features and historical sites provides a likely route.

52. Williams, *GIS Aided*, 37. See also Perttula, et al, "Caddo Ceramics from an Early 18th Century Spanish Mission in East Texas: Mission San Jose de los Nasonis (41RK200)," *Journal of Northeast Texas Archaeology* 2009 (29): 81–89. Spellings vary and include Nazones, Nasonis, and Nazonis, which is the name cited in the *Handbook of Texas*.

53. Texas General Land Office. *Spanish Land Grants*. Survey notes for the Mariano Sanchez grant in Nacogdoches County (Sanchez, Mariano; Abstract 51; File no. SC 000044:5) and Luis Sanchez (Sanchez, Luis; Abstract 51; File no. SC 000044:5) surveys are found online at the Texas General Land Office.

Chapter 3

1. Peter Cartwright and William Peter Strickland. *Autobiography of Peter Cartwright, the Backwoods Preacher*. (Cincinnati: L. Swormstedt & A. Poe for the Methodist Episcopal Church, 1856), 24-25. https://archive.org/details/autobiographyofp01cart.

2. National Park Service. "National Trails System." http://www.nps.gov/nts/nts_trails.html. Accessed November 2, 2014.

3. On old maps, the name of the trail is most often seen as Trammel, with two m's and one l. The *Handbook of Texas* spells it as such. Jack Jackson, a noted historian and Trammell descendant, made a case for Nicholas Trammell's name to be spelled with two m's and two l's, a change also reflected in the *Handbook of Texas*. This convention will be used throughout this work. Additionally, Nicholas Trammell is often referred to as Nicholas Trammell Jr., although there was no official or legal use of that designation in court or land documents. For purposes of this book, Nicholas Trammell refers to the main character of this work. Other relatives with the same name, including his father, will be identified as separate characters.

4. Family genealogies put the date of their marriage at around 1777, likely in Tennessee. The Mauldings and Trammells could have met in Virginia before their families migrated west. The State of Kentucky was formed in 1792, and Logan County was among the first counties organized soon after, making up a large portion of Western Kentucky. Logan County split and part of it became Christian County in 1796. Christian County split and part became Livingston County in 1799. Since later records place the Trammells in Livingston County, it can be assumed that they inhabited the western realm of Logan County. Morton Maulding (b. 1754), Mary (Polly) Maulding, Ambrose Maulding (b. 1755), Wesley Maulding (b. 1771), Richard Maulding (b. 1765), and Frances (Fanny) Maulding (b. 1759) were all siblings. Presley Maulding was Richard's son. Frances Maulding Trammell was Nicholas Trammell's mother.

5. W. W. Clayton. "Journal of a Voyage, Intended by God's Permission, in the Good Boat 'Adventure,' from Fort Patrick Henry, on Holston River, to the French Salt Springs, on Cumberland River, Kept by John Donelson." *History of Davidson County, Tennessee with Illustrations and Biographical Sketches of its Prominent Men and Pioneers*. (Philadelphia: J. W. Lewis & Co., 1880.)

6. Doug Drake, Jack Masters, and Bill Puryear. *Founding of the Cumberland Settlements: The First Atlas, 1779-1804*. (Gallatin, Tenn.: Warioto Press, 2009), 16-17.

7. *The Station Camp: Dogs & Deerskins*. A scene of eighteenth-century long hunters in their station camp with a half-face shelter and storage for deerskins.

8. Daniel McKinley, "A Review of the Carolina Parakeet in Tennessee," *The Migrant: A Quarterly Journal Devoted to Tennessee Birds* 50 (1979): 4.

9. Maulding's Fort was built in 1780 near the current city of Russellville, Kentucky. State Marker 1137, 10 miles south of Russellville, memorializes the location and history. Information on falls from "Seedtime," 307. Kaspar Mansker was an early settler in the region, hunting in the area as early as 1769. Mansker's Station was about twelve miles north of Nashville.

10. Harriet Louise Arnow. *Seedtime on the Cumberland.* (New York: Macmillan, 1960), 309.

11. Edward Albright. "Mansker's Party." *Early History of Middle Tennessee.* (1908.) http://www.rootsweb.com/~tnsumner/early8.htm. The numerous salt licks in the area, including the site of the French Lick, which later became the City of Nashville, were commonly named for their discoverers. Kaspar Mansker had named two licks he discovered about a mile east of Goodlettsville, the Upper and the Lower. Maulding apparently found one in between the two.

12. Nicholas Trammell's father (also named Nicholas), Ambrose Maulding, and Morton Maulding each signed the Cumberland Compact. Historians point to the pivotal role of the Cumberland Compact in helping create a governing structure for the Tennessee Territory. This agreement has been compared to the "trans-Appalachian equivalent of the Mayflower Compact." For more information see, "Cumberland Compact" in *The Tennessee Encyclopedia of History and Culture.* http://tennesseeencyclopedia.net/entry.php?rec=335.

13. John Haywood and Arthur St. Clair Colyar. *The Civil and Political History of the State of Tennessee from its Earliest Settlement up to the Year 1796.* (Nashville: W. H. Haywood, 1891), 222-23.

The date of his death can be approximated by the fact that his wife, Fanny Trammell, was petitioning the court regarding her husband's estate on January 7, 1784. See also Edythe Rucker Whitley. *Red River Settlers: Records of the Settlers of Northern Montgomery, Robertson, and Sumner Counties, Tennessee.* (Baltimore: Genealogical Publishing Company, 1980), 106.

14. Edward Albright. *Early History of Middle Tennessee.* (Nashville: Brandon Printing Company, 1909), 113-14.

15. Ibid.

16. A. W. Putnam. *History of Middle Tennessee: Or, Life and Times of Gen. James Robertson.* (Knoxville: University of Tennessee Press, 1971), 223-24.

17. From Davidson County, North Carolina (in what would become Tennessee) Court Minutes, Early Wills and Intestate Records (October 6 1783 to July 6, 1789). Abstracted from Court Minutes of Davidson County by Jeanne M. Johnson. Davidson County Clerk Minutes (Daily), Vol. M-N, 1819-1829. Microfilm roll #1602 of the Tennessee State Library & Archives, Nashville. Following the daily court records on the microfilm are pages 1-395 of the court minutes of Davidson County, North Carolina, for the period October 6, 1783 to July 6, 1789.

18. Solomon White was listed among the earliest settlers in 1780 at Nashborough. A. W. Putnam, *History of Middle Tennessee: Or, Life and Times of Gen. James Robertson,* 75-76. On July 5, 1784, White was on a Nashborough jury with Ambrose Maulding. He also received a Davidson Co. land grant in 1783. Capt. James McFadden was assigned duty on certain roads between stations near Nashville on April 1, 1783.

19. "An Inventory of the Estate of Nicholas Tramel, Deceased." *Davidson County Tennessee Will Book*, 1:10. Family History Library microfilm, 0,200,252, 1784.

20. Jack Jackson. *The Trammells: Being a History of a Pioneering Family, with Emphasis on the Descendants of Nicholas Trammell Who Blazed Trammel's Trace when Texas Was a Province of Spain*, 5. Unpublished. Copy placed with the Gonzales County Archives, Gonzales, Texas.

21. Fanny Trammell's brothers, Ambrose, Wesley (or West), and Morton held key positions in early Logan County, Kentucky. At the first meeting of the county court at Richard Maulding's house in 1792, Ambrose was elected one of the first Justices of the Peace. Wesley Maulding was Logan County's first sheriff at the same meeting and held many other elected positions. Morton Maulding was elected the first representative of Logan County to Kentucky legislature. Logan County, Ky., Order Book 1, September 25, 1792 to January 1806. September 25, 1792, 1. http://www.rootsweb.com/~kylogan/Records/Court/OB1.html.

22. From Davidson County, N.C. (later Tenn.) (Tennessee Court Minutes, Early Wills and Intestate Records, October 6 1783 to July 6, 1789). Abstracted from Court Minutes of Davidson County by Jeanne M. Johnson. Davidson County Clerk Minutes (Daily), Vol. M-N, 1819-29. Microfilm roll #1602 of the Tennessee State Library & Archives, Nashville. Following the daily court records on the microfilm are pages 1-395 of the court minutes of Davidson County, N.C., for the period October 6, 1783 to July 6, 1789

23. Logan County, Ky., Order Book 1, 25 Sept. 1792 to January 1806. March 1793, 7. http://kykinfolk.com/logan/misc/courtorderbook1.txt.

24. Logan County, Ky., Order Book 1, 25 Sept. 1792 to January 1806. 1793, 8. http://kykinfolk.com/logan/misc/courtorderbook1.txt.

25. Name sometimes seen as Askins. Zachariah Askey, Nick's stepfather, was still in Logan County in 1793, questioning why a bill of sale he was involved in was not recorded in the deed book. Logan County, Ky., Order Book 1, 25 Sept. 1792 to January 1806. March 1793, 7. http://kykinfolk.com/logan/misc/courtorderbook1.txt.

26. Shirley Wilson. *Sumner County, Tennessee Bond Book: 1787-1835*. (Hendersonville, Tenn.: Self-published, 1994). Original Bond, July Term, 1792, 53. Sumner County, Tenn.: On motion of Mrs. Askey, ordered that Phillip Trammell be appointed guardian of Nicholas Trammell, son of said Mrs. Askey, who entered into bond of 500 pounds, with David Beard and John Young, securities.

27. Willard Rouse Jillson. *The Kentucky Land Grants: A Systematic Index to All of the Land Grants Recorded in the State Land Office at Frankfort, Kentucky, 1782-1924*. (Baltimore: Genealogical Publishing Co., 1971), 420. There were as many as four Phillip Trammells who lived in the region from southern Illinois to middle Tennessee. However, this Phillip Trammell, young Nicholas's uncle, was in Logan County, Kentucky, along Red River, where he was deeded two hundred acres, Book 1, 300. Date of survey, October 7, 1797. He was still there in the 1800 census for Logan County.

28. Peter Cartwright and William Peter Strickland. *Autobiography of Peter Cartwright, the Backwoods Preacher*. (Cincinnati: L. Swormstedt and A. Poe for the

Methodist Episcopal Church, 1856), 435. https://archive.org/details/autobiographyofp01cart.

29. Cartwright, 24–25.

30. Ibid, 24.

31. Ibid, 30.

32. John T. Christian. "Part III, The Period of Growth and Organization: Chapter IV, the Great Revival of 1800," in *A History of the Baptists, Vol. II.* (New Orleans: Baptist Bible Institute, 1928.) http://www.pbministries.org/History/John%20T.%20Christian/vol2/history2_part3_04.htm.

33. Ibid.

34. Herbert Bruce Fuller. *The Purchase of Florida: Its History and Diplomacy.* (Gainesville: University of Florida Press, 1964), 141.

35. In 1776, Kentucky County, Virginia, encompassed virtually the entire state of present Kentucky. Lincoln County, Kentucky was one of three large jurisdictions formed in 1780 and included western Kentucky, where the Trammells, Mauldings, and Askeys were settling. Chickasaw land occupied the western tip of Kentucky. Logan County formed in 1792, and Mauldings were among the first elected officials in Logan County. The western half of Logan became Christian County in 1797, and Livingston County was formed from the western part of Christian County in 1799.

36. Brenda Joyce Jerome, "Livingston County, Kentucky Circuit Court Minutes 1803," *Western Kentucky Journal* XIV (2007): 24.

37. Brenda Joyce Jerome. *Livingston County, Kentucky, County Court Order Books A-B, May 1799–January 1807.* (Newburgh, Ind.: B. J. Jermoe, 1984), 21, 25.

38. Jerome, *Livingston County, Kentucky, County Court Order Books A-B, May 1799–January 1807*, 90. Livingston County was formed out of part of Logan County in 1799.

39. Jerome, *Livingston County, Kentucky, County Court Order Books A-B*, 95.

40. Jerome, *Livingston County, Kentucky, County Court Order Books A-B*, 112. Claim no. 898, on May 8, 1804.

41. Livingston County, Kentucky Tax Lists. 1805, 21. Read from microfilm at Willard Library on June 12, 2009.

42. Jerome, *Livingston County, Kentucky, County Court Order Books A-B*, 147, 153. Occurrence on October 8, 1805.

43. "Livingston County, Kentucky Tax Lists." Read from microfilm at Willard Library on June 12, 2009. Page 1. In 1808, Zachariah Askey held fifty acres of land.

44. "Livingston County, Kentucky Tax Lists." Read from microfilm at Willard Library on June 12, 2009. 1806, Lower Dist., page 1, July 6; page 18, August 11, 1807; page 1, September 11, 1807.

45. "Livingston Co, KY 1806 Delinquent Tax List." Found among loose county court papers in the Livingston County, Ky., courthouse, Smithland, Ky., by Brenda Jerome.

46. Jerome, "Livingston County, Kentucky, Circuit Court Order Book," 177. Livingston County Circuit Criminal Cases of May 1807.

47. *Nolle prosequi* is a legal term indicating that the prosecution does not wish to continue the case.

48. Ten years of settlement was required to finalize a claim. "*Pioneers and Makers of Arkansas*," Shinn, 98. *Nick's claim was proved in 1817. Proving a claim required ten years of continuous residence.* Nick Trammell, Mote Askey, and Zachariah Askey were on the earliest tax records in 1815 in Lawrence County, Mo. Other early settlers in the area included names that continued to be part of Trammell's future activities—Musick, Trimble, Burnsides, Barkman, Morrison, Magness, and Robbins were names that would continue to be seen together for many years.

49. Jackson, *Trammell's*, 5.

50. 1785, May—Heirs enter preemption cert #33. In State of N.C. (later Tenn.). 640 ac at Station Camp Creek, on Michael Shaffer's lower line.

51. Records establish that Nicholas Trammell's mother, Frances Maulding Trammell Askey, was alive in 1792 but deceased around 1800. Indications are that her death was likely in the late 1700s.

52. Six hundred forty acres in White County, Tenn., but deed entered in Franklin County. Second district, 8th range, 3rd section on Elk River.

53. This land was also referenced in a connected tract along west fork of Station Camp Creek in 1787.

54. The East Fork of Station Camp Creek begins north of Gallatin and flows southeasterly into the Cumberland River about four miles southwest of Gallatin.

55. Tennessee, William Wilcox Cooke, and Lytton Taylor. *Reports of Cases Argued and Adjudged in the Supreme Court of Errors and Appeals of Tennessee, and in the Federal Court for the District of West Tennessee.* Nashville: Printed by M. and J. Norvell, 1814. Most of the account that describes the disagreement over the land comes from the case decided by the Supreme Court of Tennessee, *John Overton v. Nicholas Trammel et al.* John Overton was a significant character in the legal and financial landscape of the state of Tennessee. Overton arrived in Nashville in 1789, where he practiced law and became a close friend of Andrew Jackson. By 1794, Overton was a land speculator on a grand scale, involving Jackson in his purchase of the land where Memphis was founded. Overton succeeded Jackson on the Superior Court in 1804.

56. Morton Maulding's daughter, Laodicea (Dicey), married Ragland Langston (her second marriage) around 1800. Dicey was Nicholas Trammell's cousin, the daughter of his uncle Morton Maulding.

57. Josiah Hazen Shinn. *Pioneers and Makers of Arkansas.* (Baltimore: Genealogical Publishing Co., 1967), 292.

58. Larry and Kathryn Priest, eds. *The Diary of Clinton Harrison Moore.* http://www.pcfa.org/genealogy/ClintonHarrisonMoore.html. Accessed December 23, 2014.

59. Livingston, Ky. Circuit Court Bundles, September 1809. Brenda Joyce Jerome, 10823. From Boynton Merrill Jr., *Jefferson's Nephews: A Frontier Tragedy.* (Lexington: University Press of Kentucky, 1987), 156.

60. *Livingston Court Circuit Court Order Book 1807–1810.* Microfilm #7012152, personal library of Brenda Joyce Jerome, 314. Action taken on Wednesday, the June 28, 1809. In a related court record, the date of May 27, 1809 was given.

61. *Livingston Court Circuit Court Order Book 1807–1810.* Microfilm #7012152, 173.

62. *Livingston Court Circuit Court Order Book 1807–1810.* Microfilm #7012152, 429.

63. Bartley Pitts is found in a Logan County, Kentucky, deed record from August 1809 as a witness to the transaction, along with Morton Maulding and Zachariah Askey (Askey). Logan County Deed Book B, 544. Pitts may have been a relative or at least a close associate of the Mauldings. They moved into Kentucky around the same time the Trammells headed west. Pitts was an heir to Morton Maulding when he died between 1818 and 1820. Trammell gave his written testimony on the matter in November 1811 and was still gathering depositions on the matter in Kentucky in 1814. Record says land was in White County, but it was actually in Franklin County. *Logan County Deed Book A*, 9. This document is signed by Nicholas Trammell and his wife Sarah.

64. William F. Pope and Dunbar H. Pope. *Early Days in Arkansas, Being for the Most Part the Personal Recollections of an Old Settler.* (Little Rock: Frederick W. Allsopp, 1895), 88.

65. *Livingston County Circuit Court Order Book, 1897–1810*, 473.

66. *Livingston County Circuit Court Order Book, 1810–1814*, 114.

Chapter 4

1. Robert Bruce Blake and Jesse J. Lee. Robert Bruce Blake Research Collection in 75 Volumes. Bexar Archives: Transcripts April 8, 1791 to January 19, 1802. Letter from Gov. J. B. Elguezabal to Jose Maria Guadiana, April 3, 1802. Vol. 23 (1958): 170.

2. David La Vere, "Between Kinship and Capitalism: French and Spanish Rivalry in the Colonial Louisiana-Texas Indian Trade," *The Journal of Southern History* 64 (May 1998): 197–218.

3. La Vere, 197–99, 203.

4. Robert B. Blake, "Sebastian Rodriguez to Gov. Antonio Cordero, February 3, 1806." Bexar Archives, Supplement V, 245.

5. La Vere, 216. See also "Kichai Indians." *Handbook of Texas Online.* June 15, 2010. Texas State Historical Association. http://www.tshaonline.org/handbook/online/articles/bmk08. Accessed December 24, 2014.

6. Francois Lagarde. *The French in Texas: History, Migration, Culture.* (Austin: University of Texas Press, 2003), 43.

7. Douglas C. McMurtrie, "First Texas Newspaper," Vol. 36, No 1, Texas State Historical Association, *The Southwestern Historical Quarterly* 36 (July 1932–April, 1933): 43. http://texashistory.unt.edu/ark:/67531/metapth101093/. Accessed December 24, 2014.

8. J. Frank Dobie. *The Mustangs.* (Boston: Little Brown and Co., 1952), 219–21.

9. Jack Jackson. *Indian Agent Peter Ellis Bean in Mexican Texas.* (College Station: Texas A&M University Press, 2005), 10. Jackson has written extensively on both Philip Nolan and Peter Ellis Bean, a key figure in Nacogdoches and Texas history who accompanied Nolan and was captured with him.

10. Dan Flores. "Bringing Home All the Pretty Horses: The Horse Trade and the Early American West, 1775–1825," in Paul Andrew Hutton, ed., *Western Heritage: A Selection of Wrangler Award-Winning Articles.* (Norman: University of Oklahoma Press,

2011). This quote is from a letter from Philip Nolan to Jesse Cook of Natchez, Mississippi on October 21, 1800.

11. Winston De Ville and Jack Jackson, "Wilderness Apollo: Louis Badin's Immortalization of the Ouachita Militia's Confrontation with the Philip Nolan Expedition of 1800," *Southwestern Historical Quarterly* XLII (January 1989): 452.

12. Blake, "Pedro de Nava to Gov. J. B. Elguezabal, May 11, 1801." Bexar Archives, Supplement IV, 116.

13. Blake, "Gov. J. B. Elguezabal to Jose Maria Guadiana, April 3, 1802." Bexar Archives, Supplement IV, 170.

14. Blake, "Letter from Nemesio Salcedo to Governor J. B. Elguezabal, May 23, 1803." Bexar Archives, Supplement IV, 204-206. From 1800 to 1813, Nemesio Salcedo was commandant general of the Interior Provinces.

15. Blake, "Letter from Nemesio Salcedo to Governor J. B. Elguezabal, March 1, 1803." Bexar Archives, Supplement IV, 191-92.

16. Blake, "Jose Miguel del Moral to Gov. J. B. Elguezabul, June 26, 1800." Bexar Archives, Supplement IV, 59.

17. Flores, 34.

18. Robert Bruce Blake. "Letter from Jose Joaquin Ugarte to Governor J. B. Elguezabal, November 4, 1804." Bexar Archives. January 5, 1799 to December 20, 1804. Supplement IV, 405.

19. Blake, "Letter from Nemesio Salcedo to William Barr, August 29, 1804." Supplement IV.

20. This location would later become known as Barr's Crossing and then Robbin's Ferry. In 1825, Nick Trammell located here and was chased off by Mexican officials in a land dispute that led to the Fredonian Rebellion.

Jean L. Epperson has thoroughly researched the old towns of Trinidad, as well as Atascocito, farther south on the Trinity River near present-day Liberty, Texas. The Spanish Royalist Army destroyed the post in September 1813. See Jean L. Epperson, *Lost Spanish Towns: Atascosito and Trinidad de Salcedo*. (Hemphill, Tex.: Dogwood Press), 1996.

21. Robert Bruce Blake. "Letter from Governor Antonio Cordero to Commandant of Nacogdoches, January 5, 1807." Robert Bruce Blake Research Collection. Supplement V, 384.

22. Flores, 173.

23. Blake, "Letter from Antonio Cordero to Nemesio Salcedo, 12/31/1805." Supplement V, 191.

24. James and Mary Dawson in the 1940s would name this section of the trail the "Spanish Trace." Trammell incorporated part of it into the road he built from Pecan Point to connect to Trammel's Trace at this point. Flores (191n) notes that Puelle's map of 1807 shows the route to the Spanish Bluff and that Anthony Glass recorded in his diary that he "crossed the road made by the Spaniards in 1807 [sic] under the command of Captain Vianne (Viana) who was in pursuit of Freeman." Two years later, Anthony Glass's party will find lingering evidence of the passage of the Spanish army in the vicinity of Naples and Douglassville, and Glass will note in his journal: "Crossed the Road made by the Spaniards in 1807 [sic] under the

command of Captain Vianne [sic] who was in pursuit of Freeman and Sparks who were ascending Red River on an exploring expedition by order of Mr. Jefferson, the Hon. President of the United States." (Glass, entry for July 12, 1808, in Flores, *Journal of an Indian Trader*, 41, 22.)

25. This point came to be called Spanish Bluff.

26. Flores, 51. Historical evidence indicates that Wilkinson was also on the Spanish payroll and created much of the unrest on the border with Spain, hoping to benefit financially. Between 1790 and 1820, there were five major Anglo-American incursions into the affairs and territory of Spain, and James Wilkinson figured in every one of them.

27. Ernest Wallace, David M. Vigness, and George B. Ward. Documents of Texas History. State House Press, 1994.

28. Robert Bruce Blake. "Letter from Pedro Lopez Prieto to Governor Manuel Salcedo, May 31, 1810." Bexar Archives, Robert Bruce Blake Research Collection, 186.

29. Robert B. Blake. "Copy of the Opinion of the Sir Military Judge," Bexar Archives, Robert Bruce Blake Research Collection. Book No. 3, Year of 1810, 88-92.

30. "El Camino del Caballo," *Handbook of Texas Online*. This road was also known as Contraband Trace and Smuggler's Road.

31. "Letter from Jose Antonio Cuellar to Pedro Lopes Prieto, January 17, 1810." Bexar Archives, Robert Bruce Blake Research Collection. Supplement VII, 25-26.

32. "Letter from Eugenio Garsilla to Pedro Lopez Prieto, May 6, 1810." Bexar Archives, Robert Bruce Blake Research Collection. Supplement VII, 90-92.

33. "Letter from Jose Agabo de Ayala to Governor Manuel Salcedo, July 13, 1810." Bexar Archives, Robert Bruce Blake Research Collection. Supplement VII, no. 202, 188.

34. Nettie Lee Benson, "Bishop Marin de Porras and Texas." *Southwestern Historical Quarterly* 50 (1947): 33.

35. "Jose Maria Guadiana to Governor Manuel Salcedo, July 22, 1810: Report of the Jurisdiction of this Village of Nacogdoches directed to the Sir Governor of the Province of Texas, Lieutenant Colonel Don Manuel de Salcedo." Bexar Archives, Blake Research Collection. September 26, 1809 to July 1, 1812, 211.

36. Epperson, 53.

37. Randell G. Tarin, ed. "The Journal of Lieutenant Colonel Don Manuel Salcedo, March 11, 1810-June 23, 1810." *The Second Flying Company of Alamo de Parras*. http://www.tamu.edu/ccbn/dewitt/adp/archives/translations/diary.html. Accessed January 2, 2015.

38. This crossing was also known as Crow's Ferry. It operated from 1796 until about 1812 and was later renamed Gaines' Ferry.

39. "Letter from Jose Miguel Crow to Governor of Texas, April 27, 1810." Bexar Archives, Blake Research Collection. September 26, 1809 to July 1, 1812. Supplement VII, 80-81.

40. Patrick G. Williams, *A Whole Country in Commotion: The Louisiana Purchase and the American Southwest*. (Fayetteville: University of Arkansas Press, 2005), 60, 73-85.

41. Julia Kathryn Garrett, "Dr. John Sibley and the Louisiana-Texas Frontier, 1803-1814 (Letter 25)," *Southwestern Historical Quarterly* XLIX (January 1946): 404.

42. Julia Kathryn Garrett, "Dr. John Sibley and the Louisiana-Texas Frontier, 1803-1814 (Letter 21)," *Southwestern Historical Quarterly* XLIX (July 1945): 116-17.

43. Black, "Pavie in the Borderlands," 28. From the Territorial Papers of the United States.

44. Fay Hempstead. *Historical Review of Arkansas: Its Commerce, Industry and Modern Affairs*. (Chicago: Lewis Publishing Company, 1911), 74.

45. Augustus Curren Jeffery and Dale Hanks. *Historical and Biographical Sketches of the Early Settlement of the Valley of White River together with a History of Izard County*. (Richmond: Jeffery Historical Society, 1973.) Originally written in 1877.

46. Stephen F. Austin used New Madrid certificates to acquire land near Fulton, Arkansas.

47. Garrett, Sibley #21, July 17, 1811, 119. Besides being home to characters labeled bad men, the settlements at Pecan Point and a little farther upriver at Jonesborough became the jumping off points for many settlers who later became part of Stephen F. Austin's Old Three Hundred original settlers. Many of these early Red River "bad men" went on to play key roles in the Republic of Texas.

48. Sibley letter to Willam Eustis, secretary of war, December 31, 1811. From Julia Kathryn Garrett. "Dr. John Sibley and the Louisiana-Texas Frontier, 1803-1814 (Letter 25)," *Southwestern Historical Quarterly* XLIX (January 1946): 403-404.

49. Flores, Freeman Custis, note on 28.

50. The Old Stone Fort has been restored and is open for viewing on the campus of Stephen F. Austin State University.

51. Epperson, 58.

Chapter 5

1. Eugene C. Barker, ed. "Austin to Bustamante, May 10, 1822." *The Austin Papers, Volume II, Part 1*. (Washington: Government Printing Office, 1924), 509.

2. The original designation of "coureur de bois" was for French traders in Illinois during the late 1700s engaging in trade with Native Americans. Lucille Fain and Harry Pettey each used it in writing articles for the Nacogdoches newspaper during the 1960s and 1970s. See Lucille Fain, "Names of Creeks Along Historic Road Are Stories in Themselves," *The Daily Sentinel*, Nacogdoches, July 9, 1963, and Harry Pettey, "Forest Couriers in Early Days Avoided Cleared Areas," *The Daily Sentinel*, Nacogdoches, May 1, 1963.

3. See Fain, Pettey. The Smuggler's Road paralleled the El Camino Real, although farther south of Nacogdoches. Historians describe its path crossing present-day State Highway 7 a little below Bernaldo Creek, passing Palisade Springs, and continuing into San Augustine County below Chireno.

4. William C. Davis. *A Way through the Wilderness: The Natchez Trace and the Civilization of the Southern Frontier*. (New York: Harper Collins, 1995), 51.

5. As a consequence of Trammell's mention in a dispute in 1813, that is the year most often given for the beginning use of Trammel's Trace. It was the year

James and Mary Dawson used in their research from the 1940s. Jack Jackson offers 1814 or 1815 as the year begun and explains that based on some unsubstantiated information, the Dawsons proposed regarding Nicholas Trammell's participation in the War of 1812.

6. Rex W. Strickland, "Miller County, Arkansas Territory, The Frontier that Men Forgot," *Chronicles of Oklahoma* 18 (March 1940): 17. http://digital.library.okstate.edu/Chronicles/v018/v018p012.html#fn7. Accessed November 20, 2004.

7. Thomas Maitland Marshall, ed. "Address of the Cherokees to his Excellency Benjamin Howard Governor, April 27, 1813," *The Life and Papers of Frederick Bates, Vol II.* (St. Louis: Missouri Historical Society, 1926): 239-41.

8. Robert A. Myers, "Cherokee Pioneers in Arkansas: The St. Francis Years, 1785-1813," *The Arkansas Historical Quarterly* 56 (1997): 153. Myers states that Connetoo was either a mixed-blood or white man adopted into the tribes. Name also seen as Coniture, Connature, and Connatine.

9. Myers, 151. The confusion of legalities around trade led to the confiscation of a large amount of goods from Connetoo not resolved until a court ruling in October 1811 forced the sale of the goods to pay off a debt Connetoo owed.

10. Myers, 153.

11. *The Missionary Herald, Containing the Proceedings at Large of the American Board of Commissioners for Foreign Missions for the Year 1829.* (Boston: Crocker and Brewster, 1829), 378. From a speech by Choctaw chief David Folsom, July 1829.

12. Shinn, 292.

13. Robert Neill. "Reminiscences of Independence County." *Publications of Arkansas Historical Association, Vol. 3.* (Little Rock: Arkansas Historical Association, 1911), 333.

14. Nicholas Trammell was sued by John M. Bradley for debt in 1828, a case that carried forward for several years. Case can be found in the collection of Territorial Briefs and Records digitized by the William H. Bowen School of Law at the University of Arkansas at Little Rock. http://arcourts.ualr.edu/case-143/143.1.htm.

15. Lovely wrote this in the margin of a letter Chisholm wrote to Agent Meigs, September 23, 1816. http://blove1014.home.comcast.net/~blove1014/Chronology.html. Accessed April 29, 2006.

16. Graves had been jailed briefly in federal prison in Little Rock in an attempt to prevent a retaliatory strike against Osage, who killed his son. He was accused of murder, but the incarceration was an attempt to derail the plan. He was later acquitted, and in an 1828 treaty Graves was awarded $1,200 "for losses sustained in his property, and for personal suffering endured by him when confined as a prisoner, on a criminal, but false accusation." http://digital.library.okstate.edu/KAPPLER/V012/treaties/che0288.htm.

17. Marshall, ed. "Address of the Cherokees," 240.

18. Grant Foreman. *Indians and Pioneers: The Story of the American Southwest Before 1830.* (New Haven: Yale University Press, 1930), 39-40. Letter from Lovely to Meigs, August 6, 1813.

19. Foreman, 51. Lovely to Governor Clark, October 11, 1814.

20. Autobiography of José Antonio Navarro, 1841, Mirabeau B. Lamar Papers, Archives and Information Services Division, Texas State Library and Archives Commission. http://www.tsl.state.tx.us/treasures/giants/navarro/navarro-auto-4.html. See also David McDonald's excellent and very complete biography of Navarro, *Jose Antonio Navarro: In Search of the American Dream in Nineteenth-Century Texas*. (Austin: Texas State Historical Association, 2010).

21. McDonald contends that Navarro was not captured by Juan Manuel Zambrano, La Bahia Presidio Captain in 1819. Zambrano was operating his own contraband operation when both he and Navarro were arrested and charged with contraband activities that year. David McDonald phone interview.

22. Foreman, 159.

23. Rex W. Strickland. "Pecan Point, TX." *Handbook of Texas Online*. June 15, 2010. Texas State Historical Association. http://www.tshaonline.org/handbook/online/articles/hrp20. Accessed January 4, 2015.

24. Foreman, 160. Quoting a letter from John Jamison, Indian agent, to the US Secretary of War, July 10, 1816.

25. Thomas Nuttall. *A Journal of Travels into the Arkansas Territory during the Year 1819*. (Fayetteville: The University of Arkansas Press, 1999), 179.

26. Ibid.

27. Jones, 323.

28. Skipper Steely. *Six Months from Tennessee: A Story of the Many Pioneers of Miller County, Arkansas*. (Wolfe City, Tex.: Henington Publishing Company, 1982). An excellent work regarding the earliest Red River settlers from Tennessee.

29. Mary Huddleston, Sammie Cantrell Rose, and Pat Taylor Wood. *Steamboats and Ferries on the White River: A Heritage Revisited*. (Conway, Ark.: UCA Press, 1998), 3-5.

30. Dewees, 11-12.

31. "Extract of a Letter from Doctor John Sibley to Doctor John H. Robinson, of this City, Dated Natchitoches, August 27 (1817)," *Natchez Intelligencer* (September 30, 1817).

32. Rex W. Strickland, "Miller County, Arkansas Territory, the Frontier that Men Forgot," *Chronicles of Oklahoma* 18 (March 1940): 27.

33. "From the New Orleans Commercial Press." *Daily National Intelligencer* V (July 23, 1817): 3.

34. Nuttall, 179.

35. Nuttall, 156, 170-71.

36. William C. Hardt and John Wesley Hardt. *Historical Atlas of Texas Methodism*. (Garland, Tex.: CrossHouse, 2008), 16-17. Stevenson's early sojourns to the lost souls on the Red River led to his being noted as Texas' first Methodist preacher.

37. Shinn, 273.

38. Trammell was on the 1816 tax list for Lawrence County.

39. Information about the names of ferry operators are from Circuit Court Records of Hempstead County.

40. "*Trammell, et al vs. Overton.*" *Middle Tennessee Supreme Court*, Box 4. Tennessee State Library and Archives, Nashville.

41. Shinn, 98.

42. When Trammell's fourth son, Henry, was born in September 1817, he was born in Kentucky. Perhaps Nicholas's wife, Sarah, and young family had not yet moved to the edge of the frontier, where trading ventures would have meant long disappearances for Nicholas, or she delivered while returning to Kentucky with her husband to settle legal matters or visit family.

43. Davis, *Natchez Trace*, 287.

44. William C. Davis. *The Pirates Laffite, The Treacherous World of the Corsairs of the Gulf*. (Orlando: Harcourt Books, 2005), 283.

45. Davis, *Lafitte*, 289.

46. Gary Pinkerton, "True Believers: Treasure Hunters at Hendricks Lake," *East Texas Historical Journal* 47 (2009): 38-47.

47. Wayne Morris, "Traders and Factories on the Arkansas Frontier, 1805-1822," *Arkansas Historical Quarterly* 28 (1969): 28.

48. Clarence Edwin Carter, ed. "John Fowler to Robert L. Coomb, Sulphur Fork Red River, Decem 29th, 1819." *The Territorial Papers of the United States, Vol XIX*. (Washington: Government Printing Office, 1953), 134. The Cherokee village Fowler references was down the Red River near Long Prairie.

49. Morris, 47.

50. Morris, 48.

51. Featherstonhaugh, Vol. II, 1-2.

52. Ralph S. Harrelson, "Two Revolutionary War Soldiers and How they Came to Hog Prairie," *Outdoor Illinois* (March 1976).

53. Jack Jackson, ed. *Almonte's Texas: Juan N. Almonte's 1834 Inspection, Secret Report & Role in the 1836 Campaign*. (Austin: Texas State Historical Association, 2003), 255. Almonte reported that citizens had abandoned the town and sought refuge in Natchitoches.

54. Dewees, 16.

55. *Arkansas Gazette* (December 11, 1819).

Chapter 6

1. Charles Summerfield. *The Rangers and Regulators of the Tanaha: Or Life Among the Lawless*. (New York: Robert M. DeWitt, 1856), viii.

2. "No. 538. Application to Make a New Selection for Pre-Emption Right, Covered by a Soldier's Bounty Right." *A Century of Lawmaking for a New Nation: U.S. Congressional Documents and Debates, 1774-1876*. American State Papers, House of Representatives, 19th Congress, 2nd Session. Public Lands, 858.

3. Independence Co Circuit Court Records. November 19, 1821. Family History Center microfilm. Jurors included William G. Shannon, Peter Tidwell, David Magness, George Gill, Daniel McNeel, Morgan Magness, Thomas Peel, Nicholas Trammil [sic], David Tidwell, John Kyler, Robert Caldwell, and James Fisher.

4. Strickland, "Miller County... Frontier Men Forgot," p. 30 for Hempstead County expansion, and creation of Miller County p. 34.

5. "Dillard, Nicholas." *Handbook of Texas Online*. June 12, 2010. Texas State Historical Association. http://www.tshaonline.org/handbook/online/articles/fdi18. Accessed January 06, 2015. Both William B. Dewees and Nicholas Dillard were part of Stephen F. Austin's Old Three Hundred colonists.

6. Dewees, 19-23.

7. Dewees, 20.

8. The Sabine River crossing of Trammel's Trace was later known as Ramsdale's Ferry or Rocky Ford. It is at the southern boundary of present Harrison County, where it connects to Rusk and Panola counties.

9. George P. Garrison, "A Memorandum of M. Austin's Journey from the Lead Mines in the County of Wythe in the State of Virginia to the Lead Mines in the Province of Louisiana West of the Mississippi, 1796-1797," *The American Historical Review* 5 (April 1900): 525-26.

10. José Erasmo Seguín to the Governor of Texas, June 23, 1821. University of Texas transcripts of the Nacogdoches Archives, January 17 to December 3, 1821. http://digital.library.okstate.edu/chronicles/v018/v018p154.html#fn30.

11. Dewees, 22.

12. Black, *Pavie in the Borderlands*, 102.

13. Stephen F. Austin, "Journal of Stephen F. Austin on his First Trip to Texas, 1821," *The Quarterly of the Texas State Historical Association* 7 (April 1904): 286-307.

14. *Austin Papers*, 508.

15. Julia Kathryn Garrett, "Dr. John Sibley and the Louisiana-Texas Frontier," *Southwestern Historical Quarterly* 49 (July 1945): 119. Letter written on July 17, 1811, from Natchitoches, La.

16. Nuttall, 170-71n.

17. Rex W. Strickland, "Chapter II: Establishment of 'Old' Miller County Arkansas Territory," *Chronicles of Oklahoma* 18 (June 1940): 157. http://digital.library.okstate.edu/Chronicles/v018/v018p154.html#fn13. Accessed January 7, 2014. Letter from William Rabb to the Governor of Texas in the summer of 1821, writing from Jonesborough on the south side of the Red River.

18. Dewees, 28. Letter IV, July 16, 1822, from Brazos River, Coahuila, and Texas.

19. Daniel Shipman. *Frontier Life: 58 Years in Texas*. (Pasadena, Tex: Abbotsford Publishing Company, 1965), 17.

20. Rex W. Strickland, "Miller County, Arkansas Territory: The Frontier that Men Forgot. Chapter III: The Final Break-up of 'Old' Miller County," *Chronicles of Oklahoma* 19 (March 1941): 44. http://digital.library.okstate.edu/Chronicles/v019/v019p037.html. Accessed January 7, 2015.

___21. Diana Everett. *The Texas Cherokees: A People Between Two Fires, 1819-1840*. (Norman: University of Oklahoma Press, 1900), 8-15.

22. *Arkansas Gazette*, 1820.

23. *Arkansas Gazette*, 1820.

24. Everett, 18-23

25. Mary Whatley Clarke. *Chief Bowles and the Texas Cherokees*. (Norman: University of Oklahoma Press, 1971), 20.

26. Clarke, 20. By reference to the "road leading from the province to the United States," the governor meant the El Camino Real.

27. Barker, S. F. Austin to Anastasio Bustamante, May 10, 1822.

28. Ibid.

29. Felipe Enrique Neri de Bastrop. "1823 Letter from Baron de Bastrop to the Political Chief of Texas Reporting on Affairs in Texas." Dorothy Sloan Books, Auction 11, Lots 1–25. December 20, 1823. Rare. Found offered for auction at $8,000–$16,000.

30. Ralph A. Wooster, ed. and Robert A. Calvert, "Stephen F. Austin," *Texas Vistas: Selections from the Southwestern Historical Quarterly* (1980).

31. Jackson, *Indian Agent*, 345.

32. W. B. Morrison, "Fort Towson," *Chronicles of Oklahoma* 8 (June 1930): 226. http://digital.library.okstate.edu/Chronicles/v008/v008p226.html. Accessed January 10, 2015.

33. Foreman, 206.

34. The route taken by the road from Pecan Point to Fort Jesup would have passed on the north and east side of the Red River in the area that was clearly US territory.

35. Josiah Shinn, *Pioneers and Makers*. Shinn estimated that four-fifths of the migration took place down the Southwest Trail. In Barker, *Colonization of Texas*, 145, Barker cites a letter of January 8, 1822, from James Bryan, Stephen F. Austin's brother-in-law, stating that "a great number will move from this state (Missouri) as from other states and the Arkansas Territory." Grant Foreman, *Indians & Pioneers*, 179, quotes an article from the 1811 *Arkansas Gazette* that "we learn from the White River that they are coming from that quarter by hundreds."

36. Rex W. Strickland, *Anglo-American Activities in Northeastern Texas, 1803–1845*. (Ph.D. diss., University of Texas, August 1937, 74). Vertical Files, Dolph Briscoe Center for American History, University of Texas at Austin. Strickland said the site was eleven miles below the site occupied by Fulton. Robert B. Musick, the trader, was living with the friendly Delaware Indians when Claiborne Wright passed by in 1816 on his way to Pecan Point.

37. "Circuit Court for Hempstead County." *Court of Common Pleas and Circuit Court, Hempstead County, Arkansas Territory, 1819–1822*. March 1822 session. Transcribed by Hempstead County Genealogical Society, 1990, 80.

38. Circuit Court for Hempstead County, 69.

39. *Arkansas Gazette* VII (March 28, 1826): whole no. 325. These were in Miller County, Arkansas, when it encompassed much of what is now northeast Texas and more of southwest Arkansas.

40. "Book C." *Court of Common Pleas and Circuit Court, Hempstead County, Arkansas Territory, Vol. 1, 1824–1828*. March 1825 session. Transcribed by Hempstead County Genealogical Society, 1993, 28. Nowlin (1798–1857) was also from Logan County, Kentucky. He was Justice of the Peace in Lafayette County. Also seen as Noland, Nowlan.

41. Court of Common Pleas, 1819–22, 59.

42. *Book C*, 49, 62.

43. Jones, 161.

44. Strickland, in his dissertation, notes a likely route from Fulton to Pecan Point was by way of Mabbitt's Salt Works (now Cerro Gordo) along the Red River on the north side. This road is described in the Order Book of the Court of Common Pleas (Hempstead County, Arkansas Territory), A, 12–13, June 10, 1819. The salt works was located in T9S, R33W, section 33.

45. Jackson, *Indian Agent*, 40. James Gaines had been commissioned by President Jefferson to survey along the Natchez Trace in 1803–1804. By 1819 he bought a long-established ferry on the El Camino Real crossing of the Sabine. He had also helped raise troops for the Gutierrez Magee Expedition. Goyens was a free man of color.

46. The Nacogdoches Archives connect Dill and Trammell a second time when on February 16, 1824, Trammell signed a second petition supporting Dill in a matter concerning Quirk's land.

47. Brenda Joyce Jerome. *Livingston County, Kentucky, Deeds Books AA-BB 1822–1829, Volume III*, 1994, 37.

48. *Austin Papers*, 705, 716.

49. Eugene C. Barker, ed. "Austin to Political Chief (March 18, 1826)." *The Austin Papers*, Vol. II, Part II, 1284. A picarro was a type of sword used in battle but was used to denote outlaws or criminals.

50. Ibid.

51. Weathers, "Town of Fulton," *Arkansas Gazette* (December 25, 1819).

52. Blake, Bexar Archives, Supplement VII, July 1, 1812 to September 7, 1824, 393.

53. In 1814, Bunch owned land in northeastern Arkansas, and in 1815, his name appeared on the probate for Thomas Howard, a member of the Lewis & Clark expedition. His name was sometimes listed as Eden Bunch or Adam Bunch.

54. Blake, Nacogdoches Archives, Book A (1800–10), Book B (1810–14), Vol. X, February 1, 1824, 321.

55. Book BB—Wills, Marriages, Pre-emptions, Estrays, etc, 1821–1845; Hempstead County, Ark. Abstracted by Hempstead County Genealogical Society, 1993, 1. Item 28, October 9, 1821.

56. *Arkansas Gazette* (January 7, 1822). Territory of Arkansas, Superior Court at October Term, 1821, *Aden Bunch vs. Bellona Bunch*, Libel for Divorce. http://www.arkansasties.com/Social/viewtopic.php?f=97&t=3150. Accessed January 11, 2015. Record blithely noted, "This day came the complainant, by his counsel, and the defendant being solemnly called, came not; and it appearing to the satisfaction of the court, that the defendant is not an inhabitant of this territory, it is ordered by the court, that publication be made in some newspaper published in this territory, six weeks in succession, and unless said defendant appear on or before the first day of our next term, and answer to the complainant's petition, the same will be taken as confessed as against her."

57. *Arkansas Gazette* (February 9, 1822). Territory of Arkansas, County of Hempstead, Court of Common Pleas, December Term, 1821. *Joshua Morrison vs. Aden Bunch.* http://www.arkansasties.com/Social/viewtopic.php?f=97&t=1491. Accessed January 11, 2015. Attachment, Oath $60.

58. Blake, Nacogdoches Archives Supplement XV, August 8, 1744, to October 26, 1825, 265-361. On the jury, there were familiar names—John York, James Dill, and Joseph Durst.

59. William Ransom Hogan, "Amusements in the Republic of Texas," *The Journal of Southern History* 3 (November 1937): 415. Material from the Nacogdoches Archives indicates that William Goyens was engaged in horse racing in 1828. Steadham's track may have been constructed as early as 1805.

According to Fain, near the Angelina River in 1876, there were still traces of a town plaza, a racetrack, and a row of house foundations. Bodan was named after a man who "loved fine horses, good whiskey, and close bets."

60. Nacogdoches Archives. Sworn Before me, the 7th of August 1824; Ellis H. Bean, alcalde.

61. The Nacogdoches Archives of September 1, 1824, noted that Nick agreed to bear costs in a suit by Stephen Lynch (Linch). The reason for the suit is not noted. It is footnoted here simply to capture the record in this account.

62. Barker, Austin Papers. John Sprowl to Austin, August 18, 1824. Sprowl was second alcalde in the Ayish Bayou district and connected to Martin Parmer in Fredonian Rebellion.

63. "*John M. Bradley v. Nicholas Trammell*, 'Territorial Briefs and Records,' University of Arkansas at Little Rock, William H. Bowen School of Law." http://arcourts.ualr.edu/about.htm. Accessed January 11, 2015. John M. Bradley was a lieutenant in the Hempstead County Militia in 1820.

64. Jack Jackson, "Nicholas Trammell's Difficulties in Mexican Texas," *East Texas Historical Journal* 38 (September 2000). There is no other time in Nick Trammell's life that is better documented than his short time in the Nacogdoches District. Following up on research done by the Dawsons during the 1940s, noted historian Jack Jackson has documented Nick's troubles in Texas by studying information available in the Blake Collection.

65. Noah Smithwick, compiled by his daughter Nanna Smithwick Donaldson. *The Evolution of a State, or Recollections of Old Texas Days*. (Austin: University of Texas Press, 1983), 1.

66. Jackson, *Indian Agent*, 32-33.

67. Blake, "Notice by Louis Procela, March 16, 1825," *Nacogdoches Archives*, Vol. LX, 301. Notice by Luis Procela. March 16, 1825.

68. Charges by Leonard Dubois made to Samuel Norris. *Nacogdoches Archives*, Vol. XXII, 319.

69. This John P. Coles and the John Cole who testified on the outcome of the horse race are apparently not the same person. John P. Coles did not live in Nacogdoches at the time. John P. Coles was part of Austin's Colony and the alcalde of the Brazos District and on the Brazos as early at 1822. He would have been familiar with any lawlessness in the colony. See Paul N. Spellman, *Old 300 Gone to Texas*. Published by Author, 2014, 243-44.

70. Letter from John P. Coles to Austin, May 13, 1825. *Austin Papers*, 1095.

71. Deposition of William Pryor before Austin, May 16, 1825. *Austin Papers*, 1095-96. Coles and Pryor were both members of the Old Three Hundred, the original settlers of Austin's colony.

72. It should be noted that several references are made to Trammell living near the Angelina, closer to Nacogdoches than his later disputed residence at the Trinity crossing of the El Camino Real. First payment was not made on that property until November 1825. Exactly where Trammell lived prior to that time is not clear. Peter Ellis Bean was also near this Angelina River location, as was a road north connecting with Trammel's Trace.

73. Testimony of Daniel O'Quinn to Samuel Norris, alcalde, April 3, 1825. Nacogdoches Archives (Blake), Vol. XI, 142.

74. Ibid.

75. Jack Jackson, *Indian Agent*, 31.

76. "*Nicholas Trammel vs. Andrew Hemphill*, Clark County Circuit Court, record 2057 (July 1825) and record 1232 (March 1828)." Special Collections of Ouachita Baptist University. http://www.obu.edu/archives/. Accessed January 23, 2015.

77. "Receipt from Haden Edwards to Nathaniel Tramel, November 26, 1825, Nacogdoches Archives, Vol. 28, page 166." Robert Bruce Black Research Collection, Vol. XV, 283.

78. James Tate's title as judge came as a result of his service in that capacity in St. Tammany Parish, Louisiana. Tate arrived in Texas sometime around 1821; his wife died that year in what is now Sabine County. One account says that Tate and his stepson, George Thomas Wharton Collins, were camping near San Augustine when Tate left nine-year-old George to go get some meat. Tate never returned. http://jliptrap.us/gen/tate.htm. Accessed January 23, 2015.

___79. Statement by Daniel O'Quinn, State of Couhuilla and Texas, District of Nacogdoches. Nacogdoches Archives, April 3, 1825, Vol. 29, 94-95. Robert Bruce Blake Research Collection, Vol. XI, 142.

80. Diary of William Fairfax Gray.

81. Eugene C. Barker. *The Life of Stephen F. Austin, Founder of Texas, 1793-1836.* (New York: Da Capo Press, 1968), 176n, 180n. Barker notes that an account given in Yoakum's *History of Texas* about the Sartuche claim ignores that it appears that Sartuche had "at least a shadow of a title" and that Edwards and Trammell were trying to push him aside. Name is sometimes seen as Sertuche or even Zertuche.

82. Robert Bruce Blake Collection, Vol. XV, Nacogdoches Archives, Book L, 1826-36, from the East Texas Research Center. November 26, 1825. See also "Government to Ignacio Sartuche, February 13, 1826." *Spanish Archives, General Land Office, Austin, TX*. Book 38, 639. From Robert Bruce Blake collection at East Texas Research Center, Nacogdoches.

83. Blake, Bexar Archives, Vol. XVI, 27.

84. Blake, Vol. XI, Nacogdoches Archive, Book D, 115. Nathaniel Trammell paid thirty dollars for surveying half a league of land on the east side of the Trinity, including the ferry, at the "St. Antone crossing."

85. Austin papers, Vol. 1, Part 2, 1383. Jack Jackson's book *Indian Agent* deals with Bean extensively. Jackson's ETHA article notes that a later Supreme Court case in 1853 demonstrated that Gaspar Flores was not authorized to grant lands either, pointing out that if Haden Edwards was illegally selling land to Trammell, so also was Flores in granting title to Sartuche. Page 29.

86. Jackson, ETHA, 31. In August, Nick's son Phillip Trammell also was summoned to Norris's court on a question of horse ownership.

87. Austin Papers, Vol. 2, 1132, Austin to Ross and Buckner, May 13, 1826.

88. Jackson, "Trammel's Difficulties," 33. He cites Blake, Red Volume XI, 246.

89. The last name is also seen as Askins or Askings.

90. Like Nick, Mote also had his own share of encounters before the court in Nacogdoches. Mote received a horse from Juan Seguin on March 13, 1824. He was ordered to appear regarding an unspecified debt to Fanny Hutchings in February 7, 1826. Askey gave testimony on a more serious matter on July 1, 1826, when he appeared in the case of some men held as traitors and counterfeiters. If anything, court records documented his lawfulness and attention to his obligations.

91. Joe E. Ericson and Carolyn R. Ericson. *Martin Parmer: The Man and the Legend.* (Nacogdoches: Ericson Books, 1999), 66. When Mote Askey was killed, it is likely his wife Lucinda (Hill) Askey was also present in Nacogdoches and would return to Arkansas with Nicholas Trammell. When she died sometime in 1831, Mote and Lucinda's son, Harrison Askey, was taken in by Nicholas Trammell and stayed with him until Harrison married in Hempstead County in 1841.

92. Louis Wiltz Kemp. *The Signers of the Texas Declaration of Independence.* (Houston: Anson Jones Press, 1944), 243-53.

93. Jackson, "Trammell's Difficulties," 33.

Chapter 7

1. Jackson, Texas by Teran, 37. Letter from General Manuel de Mier Y. Teran to President of Mexico, Guadalupe Victoria, March 28, 1828.

2. Louis J. Wortham. *A History of Texas: From Wilderness to Commonwealth*, Vol. 1. (Fort Worth: Wortham-Molyneaux Company, 1924), 213-86.

3. *Austin Papers*, 1528. Letter from Austin to Saucedo, December 4, 1826.

4. *Austin Papers*, 1558, Austin to Citizens of Victoria, January 1, 1827.

5. Robert Bruce Blake. "Yokum Gang." *Handbook of Texas Online.* June 15, 2010. Texas State Historical Association. http://www.tshaonline.org/handbook/online/articles/jey01. Accessed February 24, 2015.

6. Betje Black Klier, Pavie, 195-97.

7. "Application to Make a New Selection for Pre-Emption Right, Covered by a Soldier's Bounty Right, No. 538." A Century of Lawmaking for a New Nation: US Congressional Documents and Debates, 1774-1875, American State Papers, House of Representatives, 19th Congress, 2nd Session. Public Lands: Vol. 4, 858. http://memory.loc.gov/cgi-bin/ampage?collId=llsp&fileName=031/llsp031.db&Page=858. Accessed February 24, 2015.

8. Ibid.

9. Nicholas Trammell appeared on the tax rolls for Lafayette County, Arkansas, from 1828 to 1832. His son, Nathaniel, was on tax rolls for both Lafayette and Chicot Counties during that time period. Early Arkansas Territory Tax List—before 1830. http://files.usgwarchives.net/ar/lafayette/taxlists/laftax1.txt. Accessed February 24, 2015.

10. Robert S. Gray, ed. *A Visit to Texas: Being the Journal of a Traveler through those Parts Most Interesting to American Settlers*. (Houston: Cordova Press, 1975). Originally published in 1834.

11. Sam Williams and Mary Medearis, ed. Sam Williams, printer's devil: memorabilia: some ante-bellum reminiscences of Hempstead County, Arkansas, embracing pictures of social life, personal sketches, political annals, and anecdotes of characters and events. A. R. Hope: Etter Printing Company, 1979.

12. "Fulton: Indian's Ford and the Military Road Crossed Red River," *Arkansas Centennial* (June 1936). Collection of Southwest Arkansas Regional Archive (SARA), Washington, Ark. Acc #3773, MSF #0, V 1-11.

The cleared ground mentioned was in Township 14 S, Range 25 W, at the junction of the Spring Hill–Dooley's Ferry and the Patmos–Dooley's Ferry Road. The name "Lost Prairie" was reportedly based on the disappearance of the vast, flat prairie every time there was a flood leading any homes to find even the slightest elevation in the landscape. Buzzard Bluff on the edge of the river at the eastern side of the prairie was defined by two hills that rose like two large bubbles from the flat land surrounding them.

13. "List of Letters Remaining in the Post Office at C.H.A.T., on the 31st Dec., 1828," *Arkansas Gazette* (February 3, 1829). http://www.arkansasties.com/Social/viewtopic.php?f=97&t=5940&p=6023&hilit=+31st+Dec+#p6023. Accessed February 24, 2015. C.H.A.T. stood for Crystal Hill, Arkansas Territory, a location near Little Rock.

14. Bradley was on the township tax rolls in 1828. Bradley had been a lieutenant in the Hempstead County Militia and later was acquitted of murder.

15. "*John M. Bradley vs. Nicholas Trammel*." University of Arkansas at Little Rock, William H. Bowen School of Law, Territorial Briefs and Records. http://arcourts.ualr.edu/case-143/143.2.htm. Accessed March 15, 2012.

16. Lynn Foster. "Chester Ashley (1791–1848)." *The Encyclopedia of Arkansas History & Culture*. http://encyclopediaofarkansas.net/encyclopedia/entry-detail.aspx?entryID=1274. Accessed February 25, 2015.

17. Wavell's grant included what is now Lamar, Red River, and Bowie Counties, parts of Fannin and Hunt Counties, and Miller County, Arkansas. The Wavell Colony Register is a list of 122 families that were registered in Arthur G. Wavell's Colony during the years 1826 to 1830.

Ben Milam was killed in 1835 in the Siege of Bexar, a prelude to the Texas Revolution. Lois Garver, "Milam, Benjamin Rush," *Handbook of Texas Online*. June 15, 2010. Texas State Historical Association. http://www.tshaonline.org/handbook/online/articles/fmi03. Accessed February 25, 2015.

18. "Thursday Morning, May the 24th 1827." *Book C, Court of Common Pleas and Circuit Court, Hempstead County, Arkansas Territory*. Volume I, 1824-28. (Hope, Ark.: Hempstead County Genealogical Society, 1993), 103.

19. Lois Garver. "McKinney, Collin." *Handbook of Texas Online*. June 15, 2010. Texas State Historical Association. http://www.tshaonline.org/handbook/online/articles/fmc73. Accessed February 25, 2015. McKinney and many of his relatives signed contracts with Milam and located their new surveys by 1830-31. McKinney

was one of five delegates from Red River to the Convention of 1836 at Washington-
-on-the-Brazos.

20. *Book C*, Vol. I, 112.

21. *Book C*, Vol. I, 103-104. Although there is no land record for Martin Parmer, there is a December 1827 claim for his son, Isom, in an interesting location. Isom Parmer claimed eighty acres within Township 12 South, Range 23 West within a mile of a 160-acre claim by Nicholas Trammell.

22. *Book C*, Vol. I, 138.

23. *Book C*, Vol. I, 139.

24. *Book C*, Vol. I, 168.

25. *Book C*, Vol. I, 172.

26. *Arkansas Gazette* IX (1828): whole no. 459. List of letters remaining in the post office at Washington, Hempstead County, A.T. on the 30th day of September 1828. Letters for Martin Parmer in Chicot Co., December 1828. *Arkansas Gazette* X (1829): whole no. 474.

27. "*United States Pltt vs Isham Parmer Deft.*" *Book C, Court of Common Pleas and Circuit Court, Hempstead County, Arkansas Territory, Vol. II, 1829-1831*. (Hope, Ark.: Hempstead County Genealogical Society, 1994), 3-4.

28. *Book C*, Vol. II, 40. At the same time, the judge "ordered a scira faciesissue against Isham Parmer and Martin Parmer to show cause if any they have why an Execution should not issue against them." The correct spelling is scire facias. *Black's Law Dictionary*, seventh ed., provides the following definition: "A Writ requiring the person against whom it is issued to appear and show cause why some matter of record should not be annulled or vacated, or why a dormant judgment against that person should not be revived." Thanks to the late Judge Jim Lovett of Red River County, Texas.

29. *Book C*, Vol. II, 76-77.

30. Jackson, Teran, 4.

31. Jackson, Teran, 100.

32. Jackson, *Indian Agent*, 115.

33. Lester G. Bugbee, "Slavery in Early Texas, II," *Political Science Quarterly* 13 (December 1898): 664. The author cites the archives of the Texas State Library, nos. 328 and 326.

34. Betje Black Klier, Pavie, 199.

35. Jackson, Terán, 37. Letter from General Manuel de Mier Y Terán to president of Mexico, Guadalupe Victoria, March 28, 1828.

36. W. P. Zuber, "Thomson's Clandestine Passage around Nacogdoches," *The Quarterly of the Texas State Historical Association* 1 (1897-1898). http://texashistory.unt.edu/ark:/67531/metapth101009/. Accessed March 1, 2015.

37. The Tyler County site was about three miles west of Rockland, Texas. It was a key crossing of the Neches River at the time at an intersection of several early roads.

38. Rex W. Strickland. "Anglo-American Activities in Northeastern Texas, 1803-1845" (Ph.D. diss., University of Texas, 1937). Vertical Files, Dolph Briscoe Center for American History, University of Texas at Austin. Many thanks to Skipper Steely for a copy of this document.

39. Muriel H. Wright, "Early Navigation and Commerce along the Arkansas and Red Rivers in Oklahoma," *Chronicles of Oklahoma* 8 (1930). http://digital.library.okstate.edu/Chronicles/v008/v008p065.html. Accessed March 1, 2015. Wright quotes a Letter from Lt. James R. Stephenson to George Gibson, commissary general of subsistence, dated April 9, 1832, from Cantonment Towson, west of the Arkansas Territory, Senate Document 512, Indian Removals, Vol. I, 861–63.

40. Strickland, *Anglo-American Activities*, 181. James Ward, Jesus Morin, John Robbins, and George W. Wright were ordered to mark the road. When that group failed to complete its assignment, they were reappointed with the addition of Joseph Porter. Strickland cites George Travis Wright Papers.

41. Pat B. Clark. *The History of Clarksville and Old Red River County*. (Dallas: Mathis, Van Nort & Co, 1937), 85–86.

42. Strickland, *Anglo-American Activities*, 183. The court appointed James E. Hopkins, John Roberts, John Robbins, Henry Stout, and James Rowson to mark the road, with Hopkins and Robbins as overseers.

43. Clark, *The History of Clarksville*, 86.

44. Road was toward the home of William Gragg. *History of Clarksville*, 84. Also known as Dayton's Road. www.tshaonline.org/handbook/online/articles/DD/erd1.html. See also Strickland, 182, where he cites *Surveyor's Record book A-1 Red River County* for mention of Dayton's Road. William Gragg was one of Wavell's registered colonists who operated a sawmill near Pecan Bayou.

___45. Wright, History of Clarksville, 83.

46. James Dawson and Mary Eakin Dawson. Letter to Mr. Cora Carleton Hassford, Librarian of the Alamo. December 10, 1946. Dawson Collection at Southwest Arkansas Regional Archive (SARA) in Old Washington, Ark.

47. Dallas T. Herndon. *Centennial History of Arkansas*. (Little Rock: S. J. Clarke Publishing Company, 1922), 506–507.

48. "Southwest Trail Still Mostly Mud, Rocks, Stumps." *Events of the Early Statehood Period, 1836–1860*. http://www.oldstatehouse.com/educational_programs/classroom/arkansas_news/detail.asp?id=735&issue_id=38&page=4. Accessed March 2, 2015.

49. Mary Medearis. *Washington, Arkansas: History on the Southwest Trail*. Self-published, 1976, 17.

50. Wilford Woodruff. *Leaves from my Journal. Designed for the Instruction and Encouragement of Young Latter-Day Saints*. (Salt Lake City: Juvenile Instructor Office, 1882). http://www.cumorah.com/etexts/leavesfrommyjournal.txt. Accessed March 8, 2015.

51. These road-building specifications are not from a single source but from several sources of this time period, including Hempstead County court records and narratives.

52. Herndon, *Centennial History*, 509.

53. "Road from Jackson to Washington," *The Arkansas Gazette* 25 (June 15, 1831): column C. www.arkansasties.com. Accessed March 8, 2015.

54. Clarence Edwin Carter, "Delegate Sevier to the Secretary of War," *The Ter-*

ritorial Papers of the United States, Vol. XXI: The Territory of Arkansas, 1829–1836 (Washington: United States Government Printing Office, 1856), 900–901. Letter from Collins dated February 10, 1834.

55. Richard Peters. *The Public Statutes at Large of the United States of America: From the Organization of the Gov. in 1789, to March 3, 1845, Vol. IV.* (Boston: Little Brown & Co, 1846), 753. The full text of the statute read, "1835, Feb—An ACT to complete certain roads in the Territory of Arkansas. Be it enacted by the Senate and the House of Representatives of the United States of America, in Congress assembled, That the sum of twenty thousand dollars b, and the same is hereby, appropriated out of any money in the Treasury of the United States not otherwise appropriated, to complete the road leading from the southern boundary line of the State of Missouri, by Jackson, Little Rock, and Washington, to the town of Fulton, on the north bank of the Red River; and that the further sum of fifteen thousand dollars be appropriated in like manner to complete the military road leading from Fort Towson on the Red River, to the northern boundary line of the State of Louisiana, in the direction of Natchitoches. Approved, February 24, 1835."

56. Advertisement by William Ellis, dated October 23, 1832. *Arkansas Gazette* XIII, no. 45, whole no. 669. http://www.arkansasties.com/Social/viewtopic.php?f=97&t=8891&p=8977&hilit=+william+ellis+1832#p8977. Accessed March 8, 2015.

57. *Arkansas Gazette* (March 31, 1835). Probably meant Nacogdoches rather than Natchitoches by following that route of Trammel's Trace.

58. Dooley's Ferry was the site of later Civil War river defense structures. It remained important until replaced by a highway bridge in early 1930s.

59. These Spanish measures of land equaled together over four thousand six hundred acres.

60. Strickland, *Moscoso*, 109–37.

61. "Letter from Gideon Fitz to Elijah Hayward, Commissioner of the General Land Office, Washington City, DC. May 8, 1831." House documents, otherwise published as executive documents, 13th Congress, 2d session–49th Congress, 1st session. (Charleston: Nabu Press, 2014), 48–50.

62. "Letter from Claude A. Rankin, Commissioner, Department of State Lands, State of Arkansas to Mary Eakin Dawson, dated August 25, 1943." Dawson Collection, Southwest Arkansas Regional Archive (SARA), Old Washington, Arkansas. In the letter, Rankin notes that "Nicholas Trammell purchased the SE quarter of Section 25, Township 12S, Range 23 West (160 acres) on June 18, 1836. This quarter section is located approx four and one-half miles SSE of Emmet, one and one half miles NW of Sutton, and 3 miles S of the main Terre Rouge Creek."

63. "Letter from L. M. May, Bodcaw, Arkansas to Mr. & Mrs. J. W. Dawson, dated March 26, 1944." Vestiges of old road from Washington to Camden. Dawson Collection, Southwest Arkansas Regional Archive (SARA), Old Washington, Arkansas. May notes that the remains of an old country home that was the home of Nicholas Trammell was still visible "near the old Liberty mound (an old Indians mound)" along the road from Sutton to Emmett.

64. Medearis, *Printers Devil*, 167.

65. An *Arkansas Gazette* advertisement on May 12, 1835, notes that a man who stole twenty-five to thirty slaves has a wife living in Hempstead County near Nicholas Trammell's. http://archiver.rootsweb.ancestry.com/th/read/MAYFIELD/2002-02/1013396824. Accessed on March 8, 2015.

66. Featherstonhaugh, *Excursion*, 128.

67. *Biographical and Historical Memoirs of Southern Arkansas, Comprising a Condensed History of the State. . . Biographies of its Distinguished Citizens, a Brief Descriptive History. . . of the Counties.* (Chicago: Goodspeed Publishing Co., 1890), 555. Polly Vaughan settled on the Little Missouri River, where she ran a hotel near the later location of Jane's Ferry. Trammell is mentioned as entertaining travelers on their way to the nearby Hot Springs.

68. Hempstead County Genealogical Society. *Book BB-Wills, Marriages, Pre-emptions, Estrays, etc. 1821–1845; Hempstead County, Arkansas.* (Hope, Ark.: Hempstead County Genealogical Society, 1993), 3.

69. *Austin Papers*, 698, 798.

70. Later records in Texas establish that the Trammells operated faro games around 1853 in Fayette County, Texas.

71. Medearis, *Printers Devil*, 17.

72. Ibid.

73. Thomas Nuttall and Savoie Lottinville, ed. *A Journal of Travels into the Arkansas Territory during the Year 1819.* (Fayetteville: The University of Arkansas Press, 1999), 243.

74. Smithwick, 51–52. The "steerer" was not identified as Martin Parmer but certainly could have been based on the circumstance and location. For an excellent overview of frontier entertainment of the period, see Jodella K. Dyreson, "Sporting Activities in the American-Mexican Colonies of Texas, 1821–1835," *Journal of Sport History* 24 (1997): 269–84.

75. *Arkansas Gazette* 35 (August 18, 1835): column A.

76. *Arkansas Gazette* 35 (August 18, 1835): column C. Hempstead County Anti-Gaming Society, F. W. Campbell, chairman; Shelton Watson, secretary.

77. Ibid.

78. "Journal of the House of Representative of the United States, Vol. 2." *A Center of Lawmaking for a New Nation: U.S. Congressional Documents and Debates, 1774–1875.* American State Papers, Vol. 4, 47. http://memory.loc.gov/cgi-bin/ampage?collId=llhj&fileName=028/llhj028.db&recNum=49&itemLink=?%230280050&linkText=1. Accessed March 8, 2015.

79. Clarence Edwin Carter, ed. *The Territorial Papers of the United States, Vol. 21.* (Washington: US Government Printing Office, 1954), 1111.

80. Jackson, *Indian Agent*, 93. Jackson quotes a letter by Ben Milam. Andres Resendez discusses the market impact on situational logic applied by early immigrants in his book *Changing National Identities at the Frontier: Texas and New Mexico, 1800–1850.* (New York: Cambridge University Press, 2005), 3. The shifting loyalties of the early Pecan Point settlers is also discussed by Ted R. Worley's review of "The Territorial Papers of the United States" in *Southwestern Historical Quarterly* 60 (October 1956): 325–26.

81. Strickland, *Anglo-American Activities*, 140–41.
82. Ibid, 144.
83. Jackson, *Almonte*, 212. Almonte arrived in Nacogdoches on April 26, 1834.
84. Jackson, *Almonte*, 133.
85. Ibid, 60.
86. Jackson, *Teran*, 101.
87. Strickland, *Anglo-American Activities*, 235.
88. Jackson, *Almonte*, 143.
89. William C. Davis. *Three Roads to the Alamo: The Lives and Fortunes of David Crockett, James Bowie, and William Barret Travis*. (New York: Harper Collins, 1998), 412.
90. Davis, *Three Roads*, 412–13. The original spelling of Crockett's January 9, 1836 letter to his family is quoted.
91. Davis, *Three Roads*, 428.
92. Letter from James Bowie to Henry Rueg, Jefe Politico of the department of Nacogdoches, August 3, 1835. Texas State Library and Archives Commission, Indian Relations. http://www.tsl.state.tx.us/treasures/indians/bowie-01.html. Accessed March 11, 2015.
93. This quote is from a letter from Hempstead County, Arkansas, on December 14, 1835, written by Martin A. Poer to his father-in-law, James Poer, in Memphis. Martin A. Poer was married to Nancy Poer, born March 13, 1811, and Presley Maulding was married at Madison County, Tennessee, to her sister, Catherine Poer, who was born on September 8, 1802. The Mauldings were connected to Trammell's wife. Thanks to Bill and Wanda Trott and Frances Poer Fox.
94. Jones, *Folk Life in Early Texas*.
95. James Washington Winters, "An Account of the Battle of San Jacinto," *Quarterly of the Texas State Historical Association* 6 (October 1902): 141.

Chapter 8

1. Klier, *Pavie in the Borderlands*, xviii.
2. Resendez, 22.
3. Jackson, *Indian Agent*, 243. Letter from soldier at Nacogdoches.
4. William Fairfax Gray, *From Virginia to Texas, 1835*, 92. Diary entry for the evening of Wednesday, February 3, 1836.
5. James Weeks Tiller, "The March 11, 1837 Petition to the Congress of the Republic of Texas for the Creation of a New County Called Green," *Stirpes* 47 (2007): 5–17.
6. R. L. Jones. "Folk Life in Early Texas: The Autobiography of Andrew Davis, Part II, Cane-breaks of the Teneha," *Southwestern Historical Quarterly* 43 (1940): 327.
7. Hogan, *Amusements*, 416.
8. Walter B. Posey, "The Advance of Methodism in the Lower Southwest," *The Journal of Southern History* 2 (November 1936): 443.
9. *Arkansas Gazette* XIII (September 12, 1832): whole no. 662.
10. *Arkansas Gazette* XVII (June 14, 1836): 1027.

11. Allen D. Stokes, "Arkansas in 1836," *Phillips County Historical Quarterly* 2 (June 1964): 30.

12. *Arkansas Gazette* XVII (July 19, 1836): 1032.

13. Medearis, *Printers Devil*, 93.

14. Jones, *Autobiography of Andrew Davis*, 160.

15. Hogan, *Amusements*, 418.

16. Medearis, *Printers Devil*.

17. Hogan, *Amusements*, 31.

18. Strickland, *Anglo-American Activities*, 357-59. In 1834, Jonesborough was on the river. Strickland provides a description of the "second bottom" or second floodplain that was part of the geography bordering the Red River.

19. Ibid, 368-69.

20. *Arkansas Gazette* XVIII (October 17, 1837): whole no. 1097.

21. *Arkansas Gazette* XVIII (May 20, 1837).

22. Strickland, *Anglo-American Activities*, 359. He cites the Journal of the House of Representatives, First Congress, First Session, 252. Epperson's ferry authorization is found in Memorials and Petitions, Memorial No. 78, Box No. 26, April 1, 1837.

23. The latter was the year that his wife, Ann, died. Ramsdale said after her death that he worked about over the country, a melancholy way of saying that he had lost some of his sense of home and place. In fact, he was part of the survey crew that walked the boundary between the United States and the Republic of Texas. In the summer of 1841, they built eight-foot-tall mounds every mile along the boundary.

24. Maurice G. Fulton, ed. *Diary and Letters of Josiah Gregg: Southwestern Enterprises, 1840-1847.* Norman: University of Oklahoma Press, 1941, 117-18. See the diary entry for December 24 and 25, 1841.

25. The location of Walling's Ferry was just north of present-day Easton on a high bluff at the south bank of the Sabine River. Easton is on FM 2906, ten miles southeast of Longview at the northeastern corner of Rusk County.

26. Rusk County Deed Book A, 72.

27. Gammell's Laws of Texas, Act of January 19, 1841, Vol. 2, 538.

28. *Arkansas Gazette* XIX (February 7, 1838): whole no. 1112.

29. *Arkansas Gazette* XIX (March 4, 1838): whole no. 1118. Ad placed by J. G. and W. T. Walker, Benton County.

30. Camden, Arkansas was named Ecore a Fabre (The Bluff) until 1844.

31. *Arkansas Gazette* 180 (June 26, 1872): column A.

32. *Arkansas State Democrat* 3 (June 2, 1848): column G.

33. The historic site of Greenville is just north of the current community of Hollywood, Arkansas.

34. William Cabell Bruce. *John Randolph of Roanoke: 1773-1833, Volume II.* (London: The Knickerbocker Press, 1922), 258-60.

35. "From the Louisianian: Trade with Mexico via Red River," *New Hampshire Patriot* (August 19, 1839).

36. George Wilkins Kendall. *Narrative of the Texan Santa Fe Expedition, Volume 1.* (London: Oxford University, 1844), 121.

37. The road from Fulton was used by emigrants to California via El Paso in 1849. It used to be called the Chihuahua Trail and was renamed after 1848 to the California Trail.

38. John W. Middleton. *History of the Regulators and Moderators and the Shelby County War in 1841 and 1842, in the Republic of Texas, with Facts and Incidents in the Early History of the Republic and State, from 1837 to the Annexation, together with Incidents of Frontier Life and Indian Troubles, and the War on the Reserve in Young County in 1857.* (Fort Worth: Loving Publishing Company, 1883), 14. http://texashistory.unt.edu/ark:/67531/metapth2362/. Accessed March 21, 2015.

39. Jeffrey Williams, *GIS Aided*, 51.

40. "Central National Road." *Handbook of Texas Online*. June 12, 2010. Texas State Historical Association. http://www.tshaonline.org/handbook/online/articles/erc 01. Accessed March 21, 2015.

41. J. W. Williams and Texas State Historical Association, *The Southwestern Historical Quarterly* 47 (July 1943–April 1944). http://texashistory.unt.edu/ark:/67531/metapth146054/. Accessed March 21, 2015.

42. John Salmon Ford and Stephen B. Oates, eds. *Rip Ford's Texas*. (Austin: University of Texas Press, 1963), 27.

43. *The Telegraph & Texas Register*, reprinting an article from the November 11, 1837, *Nacogdoches Chronicle*. This story is repeated in a book by Frederic Benjamin, *Prairiedom: Rambles and Scrambles in Texas or New Estrémadura*. (New York: Paine & Burgess, 1846), 50.

44. Page, *Prairiedom*, 52–53.

45. Dorman H. Winfrey. *Texas Indian Papers, 1825–1843*. (Austin: Texas State Library, 1959), 219–21.

46. Tax schedules and census records from 1839 to 1849 place Trammell at his long-time home. Trammell is on the Hempstead County tax list for 1839.

47. This was eighty acres in the eastern half of the northeast corner of Section 36, Township 12 South, Range 23 West.

48. Trammell was on the federal census for Spring Hill Township, Hempstead County, Arkansas.

49. Betty Trammell Snyder. "Antebellum Arkansas Trammell Families: Genealogical Sketches of Trammell Families in Arkansas, 1807–1850." Unpublished and undated manuscript.

50. On the 1850 Gonzales County Free Inhabitants Schedule, Harrison is listed as thirty years old and born in Arkansas. His tombstone lists a birth date of July 15, 1820.

51. Strickland, *Anglo-American Activities*, 390–91.

52. Dawson, Trammel's Trace, SARA. Witnesses: Thomas W. Scott and J. W. Johnson. Record in Red River County, Dawson notes.

53. S. Whiting, ed, *Daily Bulletin* 1 (December 25, 1841). http://texashistory.unt.edu/ark:/67531/metapth80077/. Accessed March 21, 2015.

54. Will Book A, Lafayette County, Lewisville, Arkansas. Witnessed by R. F. Sullivan and William Trimble. From Dawson, 303.

55. Sarah Trammell died September 28, 1841.

56. Lafayette County Arkansas, Books D & E, 28, 1841, 27.

57. The land is 160 acres in the southeast quart of Section 25 in Township 12 South, Range 23 West in present-day Nevada County, Arkansas. It is just north of Highway 73, which is roughly the Old Camden to Washington Post Road and is 2.5 miles due east of the Hempstead County Line, near an ancient Indian mound.

58. Hempstead County, Arkansas. Marriages 1817–75. Hunting for bears, Arkansas marriages, 1779–1992, 2004. Original data: Arkansas marriage information taken from county courthouse records.

59. Hempstead County Court Records, 1837–46. Vol. 1. SARA. Page 227.

60. Dawson, 302. From Will Book A, Lafayette County, Arkansas.

61. *Arkansas Gazette* 15 (March 27, 1844): whole no 1265. "Horrid Murder."

62. *Houston Morning Star*, April 18, 1844.

Chapter 9

1. "A True Ghost Story." A poem by Mrs. Alta Honea in the October 30, 1941, *Nevada County Picayune*. She wrote a weekly poem for several years. Courtesy of Cathy Straley.

2. T. R. Fehrenbach. *Lone Star: A History of Texas and the Texans*. (New York: Macmillan, 1968), 281.

3. Hogan, *Amusements*, 5.

4. McClintock, "Journal of a Trip Through Texas."

5. Unknown. *Encarnacion Prisoners*. Encarnacion was a fortified hacienda, somewhat distant from other villages.

6. Ibid, 10.

7. Ibid, 10.

8. Ibid, 10.

9. Myers, *Cherokee Pioneers*, 153.

10. Unknown, *Encarnacion*, 10.

11. *Arkansas State Democrat* 21 (October 8, 1847): column C.

12. *Acts Passed at a Special Session of the General Assembly of the State of Arkansas*. (Little Rock: Woodruff and Few, 1838).

13. J. D. B. DeBow, ed. *DeBow's Review of the Southern and Western States*. (New Orleans: DeBow, 1852), 98. See also Isaac Grant Thompson, *A Practical Treatise on the Law of Highways: Including Ways, Bridges, Turnpikes and Plank Roads*. (Albany: W. C. Little, 1868).

14. DeBow, *Review*.

15. By the 1850s, there were several alternate routes from Fulton into various parts of Texas.

16. "Uses and Abuses of Lynch Law," in *The American Whig Review of Politics and Literature*. (New York: Wiley and Putnam, 1850), 460.

17. Ibid, 463.

18. *Arkansas Gazette* 9 (February 2, 1846): whole no. 1361.

19. July 13, 1846, no. 32, whole no. 1384, as a result of action in the Chicot Circuit Court, in Chancery, May Term, A.D. 1846, May 14; and also *Arkansas Gazette* 25 (May 22, 1847): whole no. 1428.

20. Henry sold his land earlier to John Trigg, in a deed witnessed by R. H. Finn (Richard Finn, Dick Finn) and Phillip Trammell. From Lafayette County, Arkansas pre-emption claim No. 4650, Patented May 1, 1845. Henry had Section 11, and Trigg owned Section 12. From Henry Trammell deed to John Trigg, February 16, 1843, Deed Book D&E, 149.

21. Henry Trammell, executor, provides power of attorney in his behalf to G. D. Royston, to transact general business. Lafayette County Courthouse. September 21, 1847, Deed Book D&E, 404.

22. Robert B. Walz, "Arkansas Slaveholdings in 1850," *Arkansas Historical Quarterly* XII (1953): 71.

23. There was a series of complicated, ongoing legal matters involving slaves given to her by her father.

24. Fayette County, civil case 659, Spring Term 1853, *Alexander Boyle v. Nathaniel Trammell*. Also *G. W. Sawyer and others, v. Alexander Boyle*, Supreme Court of Texas, 1858 (suit started January 1854).

25. Hogan, *Amusements*, 131.

26. Some of the details of this arrangement are found in civil case 659, *Alexander Boyle vs. Nathaniel Trammell*; Fayette County District Court, Spring Term 1853.

27. The Eagle Hotel was at the northeast corner of West Colorado and North Washington Streets in LaGrange, Texas. Thanks to Rox Ann Johnson.

28. *W. B. Branch vs. Nathaniel Trammell*. Fayette County (Texas) District Court, Fall Term, 1852.

29. Civil case 625, *George W. Miles vs. Nathaniel Trammell*, Fayette County (Texas) District Court, Spring Term, 1853.

30. O. C. and R. K. Hartley. "G. W. Sawyer and Others vs. Alexander Boyle." *Reports of Cases Argued and Decided in the Supreme Court of the State of Texas during the Latter Part of Galveston Term, 1858, and the Whole of Tyler Term, 1858, Vol. 21.* 1882. http://texashistory.unt.edu/ark:/67531/metapth28553/. Accessed March 21, 2015.

31. *William M. Rice vs. Nathaniel Trammel*, Fayette County (Texas) District Court, May 6, 1853.

32. O. C. Hartley and R. K. Hartley. "*Henry Trammell vs. Nicholas Trammell and Others.*" *Reports of Cases Argued and Decided in the Supreme Court of the State of Texas, during Austin Session, 1857, and Part of Galveston Session, 1858, Vol. 20.* (1882), 438. http://texashistory.unt.edu/ark:/67531/metapth28554/. Accessed March 21, 2015.

33. Henry Trammell, executor, deed to R. F. Sullivan. Lafayette County Circuit Court Records, Book D/E, March 7, 1848, 446. West half of Section 5, Township 15, Range 26.

34. *Arkansas Gazette* (January 24, 1849).

35. Ibid.

36. Hartley, *Report of Supreme Court*, Vol. 20, 418.

37. Ibid, 408.

38. Hubbard was the prosecuting attorney from 1828 to 1832 and judge of the Sixth Circuit from 1854 to 1856.

39. In 1848, Ashley County, Arkansas, was formed from part of Chicot County. Ancestry.com 1850 federal census.

40. *Texas Ranger and Lone Star* 5 (October 1, 1853).

41. *Creed Taylor v. Robert Hall and Wife.* Fayette County District Court, December 1854.

42. *The Texas Monument*, a newspaper published in La Grange, Texas, and dated April 11, 1853. Page 3, column 2.

43. Hartley, *Report of Texas Supreme Court, Vol. 20, 1858,* 409.

44. Nicholas Trammell not only sued his son, Henry, over this matter, but he also sued Harrison Askey, Mote Askey's son whom Nicholas raised.

45. Hartley, *Report of Texas Supreme Court, Vol. 20 1858,* 416.

46. *The Washington Telegraph,* July 28, 1847.

47. Joseph Barbiere. *Scraps from the Prison Table at Camp Chase and Johnson's Island.* (Doylestown, Pa: W. W. H. Davis, 1868), 241.

48. *Arkansas Gazette* 180 (June 26, 1872): column A.

49. Hempstead County Arkansas, Book H. March 19, 1849, 328. Trammell sells Becca (age twenty-one), her daughter (three), and her son (one) to Abraham Block and David Block.

50. Arkansas Census, 1819–70. Hempstead County, Canuse Township, 1850 Slave Schedule for AR. Slaves were not named but were age sixty male black, fifty-five female mulatto, forty male black, twenty-five female mulatto, twenty-five male mulatto, fifteen male black, ten male black, seven male black, three female mulatto, two male mulatto, two male mulatto.

51. *Arkansas Gazette* (August 9, 1850). Dallas County was formed in 1845, northeast of Camden.

52. Hempstead County, Book I. April 14, 1852, 207. Trammell sells eighty acres to James M. Hancock.

53. *Marks, Brands and Estrays of Hempstead County, Arkansas.* (Hope, Ark.: Hempstead County Genealogical Society, 1991), 50.

54. B Hempstead County, Book I. April 12, 1853, 543–44. Trammell sells his mark to James Hancock and sells the hogs for twenty-five bushels of corn.

55. Jackson, *The Trammells,* 50.

56. Ibid.

57. Hogan, *Amusements,* 233–34. See also Robert Thomas, *The Modern Practice of Physic, Exhibiting the Characters, Causes, Symptoms, Prognostic, Morbid Appearances, and Improved Method of Treating Diseases of all Climates,* 4th ed. (London: Longman, Hurst, Rees, Orme, and Brown, 1813), 183–84.

Chapter 10

1. From a poem by Rudyard Kipling, "The Way through the Woods."

2. "Slave Narratives, 1936." Miss Hazel Horn, the interviewer, notes that Samuels' saccount "seems far-fetched and Samuels appears to have mingled it with legends of a Spanish spying expedition headed by the pirate Jean Lafitte that explored the borderlands of Arkansas and Spanish territory around 1821. Whatever the case, the stories surrounding Samuels's grandmother have been enhanced to almost mythological proportions."

3. Medearis, *Printers Devil*, 337. Sam Williams's story of Trammell meeting Maj. Joshua Morrison on a trail is full of detail that could not have possibly been known by Williams. Sam Williams worked in the office of the Washington Telegraph in 1851. He was born in Kentucky in 1837 and came to Arkansas in 1838, moving with his family to Fulton in 1844, where his father operated taverns (Cross Keys and Union Taverns). His stories were written in 1886.

4. Ibid, 337.

5. Ibid.

6. Ibid, 337–38.

7. Ibid.

8. Bryan Erwin, "Hostelry of Yesteryear," *Nevada County Picayune* (July 1, 1934). Thanks to Cathy Straley for sharing the article.

9. Alta Honea, "A True Ghost Story," *Nevada County Picayune* (1941). Shared by Cathy Straley.

10. An article by R. P. Hamby published in the *Nevada County Picayune* on February 17, 1938, was representative of the effort. "Shortly after the arrival in this section of the Tennesseans above mentioned, a man of unsavory reputation, one Nick Trammell by name, origin and pedigree unknown, settled on Terre Rouge creek and for several years operated a tavern and gambling dive. Tradition teaches that more than one unsuspecting traveler entered Nick's hostelry leaving all hope behind and was never heard of again—suspicion being that murder was committed for then possession of whatever shekels the wayfarer possessed."

11. George Fleming Moore and Richard Sheckle Walker. "*Mary Trammell, Adm'x, v. Benjamin Shropshire and another.*" *Reports of Cases Argued and Decided in the Supreme Court of the State of Texas during Austin Session, 1858, and Part of Galveston Session, 1859. Vol. 22.* http://texashistory.unt.edu/ark:/67531/metapth28552/. Accessed March 21, 2015.

12. "*Taylor vs. Hall.*" *Reports of Cases*, Vol. 20, 1857. Harrison Askey, C. E. DeWitt, and John Watson were the last three men putting up surety bonds for Henry Trammell.

13. Frederick Law Olmstead. *The Cotton Kingdom: A Traveller's Observations on Cotton and Slavery in the American Slave States. Based upon Three Former Volumes of Journeys and Investigations.* (London: S. Low, Son & Company, 1861), 365.

14. *Galveston Daily News* (October 4, 1908).

15. Perttula, *Caddo Nation*, 12.

16. Fehrenbach, *Lone Star*, 319.

BIBLIOGRAPHY

"A Visit to Texas in 1831, Being a Journal of a Traveller Rough Those Parts Most Interesting to American Settlers with Descriptions of Scenery, Habits, Etc." *Houston Post Dispatch*, 1929.

Acts Passed at a Special Session of the General Assembly of the State of Arkansas: Which Was Begun and Held at the Capitol, in the City of Little Rock, on Monday, the Sixth Day of November, One Thousand Eight Hundred and Thirty-seven, and Ended on Monday. Little Rock: Woodruff & Pew, 1838.

Albright, Edward. "Early History of Middle Tennessee." January 1, 1908; accessed December 2014. http://www.rootsweb.ancestry.com/~tnsumner/early8.htm.

Albright, Edward. *Early History of Middle Tennessee.* Nashville: Brandon Printing, 1909.

"An Inventory of the Estate of Nicholas Tramel, Deceased." *Davidson County Tennessee Will Book* 1.10 (1784).

"Application to Make a New Selection for Pre-Emption Right, Covered by a Soldier's Bounty Right, No 538." *A Century of Lawmaking for a New Nation: U.S. Congressional Documents and Debates, 1774–1875, American State Papers, House of Representatives, 19th Congress, 2nd Session. Public Lands: Volume 4, P. 858.* December 18 1826; accessed February 24, 2015. http://memory.loc.gov/cgi-bin/ampage?collId=llsp&fileName=031/llsp031.db&Page=858.

"Arkansa Ter. Drawn and Published by F. Lucas Jr., Baltimore. B.T. Welch & Co. (1822)." *David Rumsey Historical Map Collection.* Accessed September 27, 2014. http://www.davidrumsey.com/luna/servlet/detailRUMSEY~8~1~36038~1 201240:Arkansa-Ter--Drawn-and-Published-by?qvq=q:arkansas;sort:Date;lc:R UMSEY~8~1&mi=4&trs=181.

"Arkansas Gazette Articles, 1819–1850." *Arkansas Ties.* Accessed January 11, 2015. http://www.arkansasties.com/.

Arnow, Harriette Louisa Simpson. *Seedtime on the Cumberland.* New York: Macmillan, 1960.

"Autobiography of José Antonio Navarro, 1841." *Mirabeau B. Lamar Papers.* January 1, 1841; accessed January 4, 2015. http://www.tsl.state.tx.us/treasures/giants/navarro/navarro-auto-4.html.

Austin, Stephen F., and Moses Austin. "Austin to Bustamante, May 10, 1822." In *The Austin Papers, Volume II, Part 1*, edited by Eugene C. Barker. Washington: Government. Printing Office, 1924.

Barbiere, Joseph. *Scraps from the Prison Table, at Camp Chase and Johnson's Island*. Doylestown, Pa.: W. W. H. Davis, 1868.

Barker, Eugene C. *The Life of Stephen F. Austin, Founder of Texas, 1793–1836: A Chapter of the Westward Movement by the Anglo-American People*. New York: Da Capo, 1968.

Barker, Eugene C. "Notes on the Colonization of Texas." *The Mississippi Valley Historical Review* 10, no. 2 (1923): 141–52.

Benson, Nettie Lee. "Bishop Marin De Porras and Texas." *Southwestern Historical Quarterly* 50, no. 1 (1947): 16–35.

Biographical and Historical Memoirs of Southern Arkansas, Comprising a Condensed History of the State. . . Biographies of Its Distinguished Citizens, a Brief Descriptive History . . . of the Counties . . . . Chicago: Goodspeed; 1890.

Blake, Robert Bruce. *Robert Bruce Blake Research Collection in 75 Volumes*. Vol. 23. Bexar Archives (1958): 170.

Block, W. T. *East Texas Mill Towns & Ghost Towns*. Lufkin, TX: Best of East Texas, 1994.

Bolton, Herbert Eugene. *Athanase De Mezieres and the Louisiana-Texas Frontier, 1768–1780: Documents Pub. for the First Time, from the Original Spanish and French Manuscripts, Chiefly in the Archives of Mexico and Spain*. Cleveland: Arthur H. Clark Co., 1914.

Book BB: Wills, Marriages, Estrays, Pre-emptions, Etc., Hempstead County, Arkansas, 1821–1845. Hope, Ark.: Hempstead County Genealogical Society, 1993.

Book C, Court of Common Pleas and Circuit Court (Vol. 1, 1824–1828), Hempstead County, Arkansas Territory, 1819–1822. Transcribed by Hempstead County Genealogical Society, 1993.

Book C, Court of Common Pleas and Circuit Court (Vol. II, 1829–1831). Hope, Ark.: Hempstead County Genealogical Society, 1994.

Book C, Court of Common Pleas and Circuit Court, Hempstead County, Arkansas Territory (Vol. 1, 1824–1828). Hope, Ark.: Hempstead County Genealogical Society, 1993.

Bruce, William Cabell. *John Randolph of Roanoke, 1773–1833: A Biography Based Largely on New Material, Vol. II*. London: Knickerbocker, 1922.

Buckley, Samuel Botsford. *Geological and Agricultural Survey of Texas*. Houston: A. C. Gray, State Printer, 1874.

Bugbee, Lester G. "Slavery in Early Texas II." *Political Science Quarterly* 13.4 (1898): 664.

Carter, Clarence Edwin. *The Territorial Papers of the United States, Vol. XXI*. Washington: United States Government Printing Office, 1934.

Carter, Clarence Edwin, ed. *The Territorial Papers of the United States, Vol. XIX*. Washington: United States Government Printing Office, 1953. Letter from John Fowler to Robert L. Coomb, Sulphur Fork of Red River, December 28, 1819.

Cartwright, Peter, and William P. Strickland. *Autobiography of Peter Cartwright, the Backwoods Preacher.* January 1 1856; accessed December 23, 2014. https://archive.org/details/autobiographyofp01cart.

"Central National Road." *Handbook of Texas Online.* Texas State Historical Association. June 12, 2010; accessed March 21, 2015. https://tshaonline.org/handbook/online/articles/erc01.

Christian, John T. "A History of the Baptists, Volume II," in *The Great Revival of 1800.* Baptist History Throughout the Ages. http://pbministries.org/History/John%20T.%20Christian/v0l2/index.htm.

Clark, Pat B. *The History of Clarksville and Old Red River County.* Dallas: Mathis, Van Nort, 1937.

Clarke, Mary Whatley. *Chief Bowles and the Texas Cherokees.* Norman: University of Oklahoma, 1971.

"Journal of a Voyage, Intended by God's Permission, in the Good Boat 'Adventure,' from Port Patrick Henry, on Holston River, to the French Salt Springs on Cumberland River, Kept by John Donelson." In *History of Davidson County, Tennessee with Illustrations and Biographical Sketches of Its Prominent Men and Pioneers,* edited by. W. W. Clayton. Philadelphia: J. W. Lewis, 1880.

Court of Common Pleas and Circuit Court, Hempstead County, Arkansas Territory, 1819–1822. Arkansas: Transcribed by Hempstead County Genealogical Society, 1990.

Davis, William C. *A Way through the Wilderness: The Natchez Trace and the Civilization of the Southern Frontier.* New York: HarperCollins, 1995.

Davis, William C. *A Way through the Wilderness: The Natchez Trace and the Civilization of the Southern Frontier.* New York: HarperCollins, 1995.

Davis, William C. *Three Roads to the Alamo: The Lives and Fortunes of David Crockett, James Bowie, and William Barret Travis.* New York: Harper Collins Publishers, 1998.

Davis, William C. *The Pirates Laffite: The Treacherous World of the Corsairs of the Gulf.* Orlando: Harcourt, 2005.

Dawson, James, and Mary Eakin Dawson. "Letter to Mrs. Cora Carleton Hassford, Librarian of the Alamo." *Dawson Collection.* December 10, 1947.

Dawson, James, and Mary Eakin Dawson. *Trammel's Trace.* Washington, Ark.: Unpublished, 1944. Manuscripts and research maintained in the Southwest Arkansas Regional Archive (SARA).

De Bastrop, Felipe Enrique Neri. "1823 Letter from Baron De Bastrop to the Political Chief of Texas Reporting on Affairs in Texas." *Dorothy Sloan Books, Auction 11, Lots 1–25.* December 20, 1823; accessed January 10, 2015. http://www.dsloan.com/Auctions/A11/001-025Web.htm.

De Ville, Winston, and Jack Jackson. "Wilderness Apollo: Louis Badin's Immortalization of the Ouachita Militia's Confrontation with the Philip Nolan Expedition of 1800." *Southwestern Historical Quarterly* XLII.3 (1989).

DeBow, J. D. B., ed. *DeBow's Review of the Southern and Western States.* New Orleans: DeBow, 1852.

DeWees, William Bluford, and Cara Cardelle. *Letters from an Early Settler of Texas.* Louisville: Morton & Griswold, 1852.

Dobie, J. Frank, and Charles Banks Wilson. *The Mustangs.* Boston: Little Brown, 1952.

Drake, Douglas, and Jack Masters. *Founding of the Cumberland Settlements: The First Atlas, 1779–1804, Vol. I.* Gallatin: Warioto Press, 2009.

"El Camino Del Caballo." *Handbook of Texas Online.* Texas State Historical Association. Accessed January 2, 2015. http://www.tshaonline.org/handbook/online/articles/exehh.

Ellis, William. "New Ferry at Point Remove." *Arkansas Ties: Arkansas Gazette 1819–1930.* October 23, 1832; accessed March 8, 2015. http://www.arkansasties.com/Social/viewtopic.php?f=97&t=8891&p=8977&hilit=williamellis1832#p8977.

Encarnacion Prisoners Comprising an Account of the March of the Kentucky Cavalry from Louisville to the Rio Grande: Together with an Authentic History of the Captivity of the American Prisoners, including Incidents and Sketches of Men and Things on the R. Louisville: Prentice and Weissinger, 1848.

Epperson, Jean L. *Lost Spanish Towns: Atascosito and Trinidad De Salcedo.* Hemphill, Tex.: Dogwood Press, 1996.

Eugen, Barker, and Eugene C. Barker. *The Austin Papers.* Washington: Government Printing Office, 1924.

Everett, Dianna. *The Texas Cherokees: A People between Two Fires, 1819–1840.* Norman: University of Oklahoma, 1990.

Fain, Lucille. "Names of Creeks Along Historic Road Are Stories in Themselves." *The Daily Sentinel.* July 9, 1963.

Featherstonhaugh, George William. *Excursion through the Slave States, from Washington on the Potomac to the Frontier of Mexico; With Sketches of Popular Manners and Geological Notices.* New York: Harper, 1844.

Fehrenbach, T. R. *Lone Star: A History of Texas and the Texans.* New York: Macmillan, 1968.

Fieth, Kenneth. *Cumberland Compact.* December 25, 2009; accessed December 22, 2014. http://tennesseeencyclopedia.net/entry.php?rec=335.

Flores, Dan. "Bringing Home All the Pretty Horses: The Horse Trade and the Early American West, 1775–1825" In *Western Heritage: A Selection of Wrangler Award-Winning Articles,* edited by Paul Andrew Hutton. Norman: University of Oklahoma, 2011.

Flores, Dan L. *Southern Counterpart to Lewis & Clark: The Freeman & Custis Expedition of 1806.* Norman: University of Oklahoma, 2002.

Ford, John Salmon, and Stephen B. Oates. *Rip Ford's Texas.* Austin: University of Texas, 1963.

Foreman, Grant. *Indians & Pioneers: The Story of the American Southwest before 1830.* New Haven: Yale University Press, 1930. Letter from Indian Agent Lovely to Agent Meigs, August 6, 1813.

Foster, Lynn. "Chester Ashley (1791–1848)." *The Encyclopedia of Arkansas History & Culture.* September 24, 2012; accessed February 26, 2015. http://www.ency

clopediaofarkansas.net/encyclopedia/entry-detail.aspx?entryID=1274.
"From the New Orleans Commercial Press." *Daily National Intelligencer.* July 23, 1817.
Fuller, Hubert Bruce. *The Purchase of Florida, Its History and Diplomacy.* Gainesville: University of Florida, 1964.
Garrett, Julia Kathryn. "Dr. John Sibley and the Louisiana-Texas Frontier, 1803-1814 (Letter 25)." *Southwestern Historical Quarterly* 49, no. 1 (1946): 44.
Gregg, Josiah. *Diary & Letters of Josiah Gregg: Southwestern Enterprises, 1840-1847,* edited by Maurice G. Fulton. Norman: University of Oklahoma, 1941.
Garrett, Julia Kathryn. "Dr. John Sibley and the Louisiana-Texas Frontier." *Southwestern Historical Quarterly* 49.1 (1945): 116-19.
Garrison, George P. "A Memorandum of M. Austin's Journey from the Lead Mines in the County of Wythe in the State of Virginia to the Lead Mines in the Province of Louisiana West of the Mississippi, 1796-1797." *The American Historical Review* 5.3 (1900): 518-23.
Garver, Lois. "McKinney, Collin." *Handbook of Texas Online.* Texas State Historical Association. June 15, 2010; accessed February 26, 2015. http://www.tshaonline.org/handbook/online/articles/fmc73.
Garver, Lois. "Milam, Benjamin Rush." *Handbook of Texas Online.* Texas State Historical Association. June 15, 2010; accessed February 26, 2015. http://www.tshaonline.org/handbook/online/articles/fmi03.
Gray, Robert S., ed. *A Visit to Texas in 1831, Being the Journal of a Traveller through Those Parts Most Interesting to American Settlers, with Descriptions of Scenery, Habits, Etc.* Houston: Cordovan, 1975.
Gray, William Fairfax. *From Virginia to Texas, 1835: Diary of Col. Wm. F. Gray, Giving Details of His Journey to Texas and Return in 1835-1836 and Second Journey to Texas in 1837.* Houston: Fletcher Young, 1965. Reprint.
Gregg, Josiah. *Commerce of the Prairies, Or, The Journal of a Santa Fè Trader: During Eight Expeditions across the Great Western Prairies, and a Residence of Nearly Nine Years in Northern Mexico.* New York: H. G. Langley, 1844.
Hardt, William C., and John Wesley Hardt. *Historical Atlas of Texas Methodism.* Garland, Tex.: CrossHouse, 2008.
Harrelson, Ralph S. "Two Revolutionary War Soldiers and How They Came to Hog Prairie." *Outdoor Illinois.* March 1, 1976.
"Reports of Cases Argued and Decided in the Supreme Court of the State of Texas, during Austin Session, 1857, and Part of Galveston Session, 1858. Volume 20." In *Reports of Cases Argued and Decided in the Supreme Court of the State of Texas, during Austin Session, 1857, and Part of Galveston Session, 1858. Volume 20, 1882,* edited by O. C. Hartley and R. K. Hartley. Supreme Court of Texas. January 1, 1858.
"Reports of Cases Argued and Decided in the Supreme Court of the State of Texas during the Latter Part of Galveston Term, 1858, and the Whole of Tyler Term, 1858. Volume 21." In *The Portal to Texas History,* edited by O. C. Hartley and R. K. Harley. January 1, 1859.

Haywood, John. "The Civil and Political History of the State of Tennessee from Its Earliest Settlement Up to the Year 1796." *The Civil and Political History of the State of Tennessee from Its Earliest Settlement Up to the Year 1796.* January 1, 1891.

Hempstead, Fay. *Historical Review of Arkansas: Its Commerce, Industry and Modern Affairs.* Chicago: Lewis, 1911.

Hempstead, Fay. *Historical Review of Arkansas: Its Commerce, Industry and Modern Affairs.* Greenville, S.C.: Southern Historical Press, 1978. Reprint.

Herndon, Dallas T. *Centennial History of Arkansas, Vol. 1.* Little Rock: S. J. Clarke, 1922.

Hogan, William Ransom. "Amusements in the Republic of Texas." *The Journal of Southern History* 3.4 (1937): 397–421.

"Letter from Gideon Fitz to Elijah Hayward, Commissioner of the General Land Office, Washington City, DC. May 8, 1831." *House Documents, Otherwise Publ. as Executive Documents, 13th Congress, 2d Session-49th Congress, 1st Session.* Charleston: Nabu, 2014.

Jackson, Jack, editor, and John Wheat, translator. *Texas by Terán: The Diary Kept By General Manuel De Mier Y Terán On His 1828 Inspection of Texas.* Austin: University of Texas, 2000.

Jackson, Jack. *The Trammells: Being a History of a Pioneering Family, With Emphasis on the Descendants of Nicholas Trammell Who Blazed Trammel's Trace When Texas Was a Province of Spain.* Gonzales, Tex.: Unpublished, Copy with Gonzales County Archives, 1999.

Wheat, John. *Almonte's Texas: Juan N. Almonte's 1834 Inspection, Secret Report, and Role in the 1836 Campaign,* edited by Jack Jackson. Austin: Texas State Historical Association, 2003.

Jackson, Jack. *Indian Agent Peter Ellis Bean in Mexican Texas.* College Station: Texas A&M University Press, 2005.

Jackson, Jack. "Nicholas Trammell's Difficulties in Mexican Texas." *East Texas Historical Journal* 38, no. 2 (2000): 15–39.

"James Bowie, Report to Political Chief, 1835." *Texas State Library and Archives Commission, Indian Relations.* August 3, 1835.

Jeffery, Augustus C. *Historical and Biographical Sketches of the Early Settlement of the Valley of White River Together with a History of Izard County.* Richmond: Jeffery Historical Society, 1973.

Jerome, Brenda Joyce. "Livingston County, Kentucky Circuit Court Minutes 1803." *Western Kentucky Journal* XIV.4 (2007): 24.

Jerome, Brenda Joyce. *Livingston County, Kentucky, Deeds Books AA-BB, 1822–1829, Vol 1.* Newburgh, Ind: B. J. Jerome, 1994.

Jerome, Brenda Joyce. *Livingston County, Kentucky, County Court Order Books A-B, May 1799-January 1807.* Newburgh, Ind: B. J. Jerome, 1994.

Jillson, Willard Rouse. *The Kentucky Land Grants; a Systematic Index to All of the Land Grants Recorded in the State Land Office at Frankfort, Kentucky, 1782–1924.* Baltimore: Genealogical Pub., 1971.

"John M. Bradley vs. Nicholas Trammel." University of Arkansas at Little Rock, William

H. Bowen School of Law, Territorial Briefs and Records. June 28, 1828; accessed March 15, 2012. http://arcourts.ualr.edu/case-143/143.2.htm.

Jones, R. L. "Folk Life in Early Texas: The Autobiography of Andrew Davis." *The Southwestern Historical Quarterly* 43.2 (1940): 161. http://texashistory.unt.edu/ark:/67531/metapth101111/.

Jones, R. L. "Folk Life in Early Texas: The Autobiography of Andrew Davis, Part II: Cane-breaks of the Teneha." *Southwestern Historical Quarterly* 43.3 (1940): 323-41.

"Journal of the House of Representatives of The United States, Vol. 2." *A Century of Lawmaking for a New Nation: U.S. Congressional Documents and Debates, 1774–1875, American State Papers, House of Representatives, 19th Congress, 2nd Session. Public Lands: Volume 4, P. 858.* December 9, 1834; accessed March 10, 2015. http://memory.loc.gov/cgi-bin/ampage?collId=llhj&fileName=028/llhj028.db&recNum=49&itemLink=?#0280050&linkText=1.

Kemp, Louis Wiltz. *The Signers of the Texas Declaration of Independence.* Houston: Anson Jones Press, 1944.

Kendall, Geo. Wilkins. *Narrative of the Texan Santa Fé Expedition: Comprising a Description of a Tour through Texas and across the Great Southwestern Prairies the Camanche and Caygüa Hunting-grounds: With an Account of the Sufferings from Want of Food, Losses from Hostile In, Vol. 1.* London: Oxford University, 1844.

Kendall, George Wilkins. *Narrative of the Texan Santa Fe Expedition Comprising a Tour through Texas and Capture of the Texans.* London: Wiley & Putnam, 1844.

"Kichai Indians." *Handbook of Texas Online.* Texas State Historical Association. 2010. https://tshaonline.org/handbook/online/articles/bmk08.

Klier, Betje Black, and Théodore Pavie. *Tales of the Sabine Borderlands: Early Louisiana and Texas Fiction.* College Station: Texas A&M University Press, 1998.

Klier, Betje Black, and Théodore Pavie. *Pavie in the Borderlands: The Journey of Théodore Pavie to Louisiana and Texas, 1829–1830, including Portions of His Souvenirs Atlantiques.* Baton Rouge: Louisiana State University Press, 2000.

La Vere, David. "Between Kinship and Capitalism: French and Spanish Rivalry in the Colonial Louisiana-Texas Indian Trade." *The Journal of Southern History* 64.2 (1998): 197-99, 203.

"Lafayette CO. Early Arkansas Territory Tax List—Before 1830." *US GenWeb Archives.* Accessed February 25, 2015. http://files.usgwarchives.net/ar/lafayette/taxlists/laftax1.txt.

Lagarde, Francois. *The French in Texas History, Migration, Culture.* Austin: University of Texas, 2003.

"List of Letters Remaining in the Post Office at C.H.A.T., on the 31st Dec., 1828." *Arkansas Ties.* Accessed February 25, 2015. http://www.arkansasties.com/Social/viewtopic.php?f=97&t=5940&p=6023&hilit= 31st Dec #p6023.

Lyne, Judy Utley. *Logan County, KY, Order Book, 25 Sept 1792 to January 1806, Pg 7.* January 1, 1793; accessed December 22, 2014. http://kykinfolk.com/logan/misc/courtorderbook1.txt.

"Maps." *American Memory Project.* Library of Congress. Accessed September 27, 2014. http://memory.loc.gov/ammem/browse/ListSome.php?category=Maps.

Marks, Brands, and Estrays of Hempstead County, Arkansas, 1819–1959. Hope, Ark.: Hempstead County Genealogical Society, 1991.

Marryat, Frederick. *The Travels and Adventures of Monsieur Violet: In California, Sonora, and Western Texas*. New York: George Routledge and Sons, 1843.

"Address of the Cherokees to His Excellency Benjamin Howard Governor, April 27, 1813." In *The Life and Papers of Frederick Bates, Vol. II*, edited by Thomas Maitland Marshall. St. Louis: Missouri Historical Society, 1926.

Martin, James C., and Robert Sidney Martin. *Maps of Texas and the Southwest, 1513–1900*. Austin: Texas State Historical Association, 1999.

McClintock, William. "Journal of a Trip Through Texas and Northern Mexico in 1846–1847." *Southwestern Historical Quarterly* 34.1 (1931). http://texashistory.unt.edu/ark:/67531/metapth101091/m1/24/?q=april 1931.

McDonald, David. *Jose Antonio Navarro: In Search of the American Dream in Nineteenth Century Texas*. Austin: Texas State Historical Association, 2010.

McKinley, Daniel. "A Review of the Carolina Parakeet in Tennessee." *The Migrant: A Quarterly Journal Devoted to Tennessee Birds* 50.1 (1979): 4.

McMurtrie, Douglas C. "First Texas Newspaper." *Southwestern Historical Quarterly* 36.1 (1933): 43. http://texashistory.unt.edu/ark:/67531/metapth101093/m1/49/.

Medearis, Mary. *Washington, Arkansas: History on the Southwest Trail*. Washington, Ark.: Medearis, 1976.

Williams, Sam. *Sam Williams, Printer's Devil: Memorabilia: Some Ante-bellum Reminiscences of Hempstead County, Arkansas Embracing Pictures of Social Life, Personal Sketches, Political Annals, and Anecdotes of Characters and Events*, edited by Mary Medearis. Hope, Ark.: Etter Printing, 1979.

Middleton, John W. *History of the Regulators and Moderators and the Shelby County War in 1841 and 1842, in the Republic of Texas with Facts and Incidents in the Early History of the Republic and State, from 1837 to the Annexation, Together with Incidents of Frontier Life an*. Fort Worth: Loving Pub., 1883.

"Reports of Cases Argued and Decided in the Supreme Court of the State of Texas during Austin Session, 1858, and Part of Galveston Session, 1859. Volume 22." In *The Portal to Texas History*, edited by George Fleming Moore and Richard Sheckle Walker. Supreme Court of Texas. January 1, 1859; edited March 21, 2015.

Morris, Wayne. "Traders and Factories on the Arkansas Frontier, 1805–1822." *Arkansas Historical Quarterly* 28 (1969): 28–48.

Myers, Robert A. "Cherokee Pioneers in Arkansas: The St. Francis Years, 1785–1813." *The Arkansas Historical Quarterly* 56.2 (1997): 127–57.

Neill, Robert. "Reminiscences of Independence County." *Publications of Arkansas Historical Association, Vol. 3*. Little Rock: Arkansas Historical Association, 1911. 333.

"*Nicholas Trammell vs. Andrew Hemphill*, Clark County Arkansas Circuit Court Records." *Special Collections of Ouachita Baptist University*. January 1, 1825.

Norvell, Helen P. *King's Highway, the Great Strategic Military Highway of America, El Camino Real, the Old San Antonio Road*. Austin: Firm Foundation Pub. House, 1945.

Nuttall, Thomas. *A Journal of Travels into the Arkansas Territory during the Year 1819*. Fayetteville: University of Arkansas, 1999.

Nuttall, Thomas. *A Journal of Travels into the Arkansas Territory, during the Year 1819*, edited by Savoie Lottinville. Fayetteville: University of Arkansas, 1999.

Oliphant, Jimmy, editor. *Jeremy Hilliard: Gone to Texas*. Marshall, Tex.: Self-published.

Olmstead, Frederick Law. "The Cotton Kingdom: A Traveller's Observations on Cotton and Slavery in the American Slave States. Based upon Three Former Volumes of Journeys and Investigations." London: S. Low, Son & Company, 1861.

Page, Frederic Benjamin. *Prairiedom Rambles and Scrambles in Texas or New Estrémadura*. New York: Paine & Burgess, 1846.

Perttula, Timothy K. *The Caddo Nation: Archaeological and Ethnohistoric Perspectives*. Austin: University of Texas, 1992.

Perttula, Timothy K., Bill Young, Paul Shawn Marceaux. "Caddo Ceramics from an Early 18th Century Spanish Mission in East Texas: Mission San Jose de los Nasonis (41RK200)." *Journal of Northeast Texas Archaeology* 29 (2009): 81–89.

Peters, Richard. *The Public Statutes at Large of the United States of America, from the Organization of the Government in 1789, to March 3, 1845, Vol. VIV*. Boston: C. C. Little and J. Brown, 1845.

Pettey, Harry. "Forest Couriers in Early Days Avoided Cleared Areas." *The Daily Sentinel*. May 1, 1964.

Pinkerton, Gary. "True Believers: Treasure Hunters at Hendricks Lake." *East Texas Historical Journal* 47.2 (2009): 38–47.

Pope, William F., and Dunbar H. Pope. *Early Days in Arkansas Being for the Most Part the Personal Recollections of an Old Settler*. Little Rock: Frederick W. Allsopp, 1895.

Posey, Walter B. "The Advance of Methodism into the Lower Southwest." *The Journal of Southern History* 2.4 (1936): 439–52.

"The Dairy of Clinton Harrison Moore," edited by Larry and Kathryn Priest. January 1, 1904; accessed December 23, 2014. http://www.pcfa.org/genealogy/ClintonHarrisonMoore.html.

Priest, Larry, ed., and Kathryn Priest. "The Diary of Clinton Harrison Moore." *Prescott Community Freenet Association*. Accessed September 27, 2014. http://www.pcfa.org/genealogy/ClintonHarrisonMoore.html.

Putnam, A. W. *History of Middle Tennessee; Or, Life and times of Gen. James Robertson*. Knoxville: University of Tennessee, 1971.

Resendez, Andre. *Changing National Identities at the Frontier: Texas and New Mexico, 1800–1850*. Cambridge, U.K.: Cambridge University Press, 2005.

"Road from Jackson to Washington." *Arkansas Gazette* 2 (1831): column C.

"Rusk County Sketch File 10." *Historical Map Collection*. Texas General Land Office. Accessed September 27, 2014. http://www.glo.texas.gov/history/archives/map-store/index.cfm#item/35507.

Shinn, Josiah Hazen. *Pioneers and Makers of Arkansas*. Baltimore: Genealogical, 1967.

Shipman, Daniel. *Frontier Life: 58 Years in Texas.* Pasadena, Tex.: Abbotsford Pub., 1965.

"Slave Narratives: A Folk History of Slavery in the United States from Interviews with Former Slaves." *The Federal Writers Project 1936–1938.* Library of Congress, Public Works Administration. January 1, 1941.

Smithwick, Noah, and Nanna Smithwick Donaldson. *The Evolution of a State, Or, Recollections of Old Texas Days.* Austin: University of Texas, 1983.

"Southwest Trail Home Page." *Arkansas Southwest Trail Research.* Accessed September 27, 2014. http://www.southwesttrail.com.

"Southwest Trail Still Mostly Mud, Rock, Stumps." *Events of the Early Statehood Period, 1836–1860.* January 1, 1834; accessed March 2, 2015. http://www.oldstatehouse.com/collections/classroom/arkansas_news.aspx?issue=38&page=4&detail=735.

"Spanish Land Grants." *GLO Historical Collections.* Texas General Land Office. September 27, 2014. http://www.glo.texas.gov/cf/land-grant-search/LandGrants Worklist.cfm?PageNum_qrylandgrants=3.

Spellman, Paul N. *Old 300: Gone to Teas.* Self-Published by Author, 2014.

Stacey, Truman. "Athanase De Mezieres." *Notable Men and Women of Louisiana.* Southwest Louisiana Historical Association. Accessed September 2014. http://www.swlahistory.org/mezieres.htm.

Steely, Skipper. *Six Months from Tennessee: A Story of the Many Pioneers of Miller County, Arkansas . . ."* Wolfe City, Tex.: Henington Pub., 1982.

Stokes, Allen D. "Arkansas in 1836." *Phillips County Historical Quarterly* 2.4 (1964): 28–32.

Strickland, Rex W. "Moscoso's Journey through Texas." *Southwestern Historical Quarterly* 45.2 (1942): 109–37.

Strickland, Rex W. "Miller County, Arkansas Territory, The Frontier That Men Forgot." *Chronicles of Oklahoma* 18, no. 1 (1940). Accessed January 4, 2015. http://digital.library.okstate.edu/Chronicles/v018/v018p012.html#fn7.

Strickland, Rex W. "Miller County, Arkansas Territory: The Frontier that Men Forgot, Chapter III: The Final Break-up of 'Old' Miller County." *Chronicles of Oklahoma* 19, no. 1 (1941). Accessed January 8, 2015. http://digital.library.okstate.edu/Chronicles/v019/v019p037.html.

Strickland, Rex W. *Anglo-American Activities in Northeastern Texas, 1803–1845.* Ph.D. diss., 1936.

"Tate." *Jim Liptrap's Homepage.* Accessed January 25, 2015. http://jliptrap.us/gen/tate.htm.

"Tejas: Life and Times of the Caddo." *Tejas Main.* University of Texas at Austin, College of Liberal Arts. Accessed September 27, 2014. http://www.texasbeyondhistory.net/tejas/index.html.

"The Journal of Manuel Salcedo." The Second Flying Company of Alamo De Paras. January 1, 1810; accessed January 2, 2015. http://www.tamu.edu/ccbn/dewitt/adp/archives/translations/diary.html.

"A Speech by Choctaw Chief, David Folsom." In *The Missionary Herald, Containing the Proceedings At Large of the American Board of Commissioners for Foreign Missions for the Year 1829,* 378. Boston: Crocker and Brewster, 1829.

Thomas, Robert. *The Modern Practice of Physic: Exhibiting the Characters, Causes, Symptoms, Prognostic, Morbid Appearances, and Improved Method of Treating the Diseases of All Climates.* London: Longman, Hurst, Rees, Orme, and Brown..., 1813.

Thompson, Isaac Grant. *A Practical Treatise on the Law of Highways including Ways, Bridges, Turnpikes, and Plank Roads, at Common Law and under the Statutes: With an Appendix of Forms.* Albany: W. C. Little, 1868.

Tiller, James Weeks. "The March 11, 1837 Petition to the Congress of the Republic of Texas for the Creation of a New County Called Green." *Stirpes* 47 (2007): 5-17.

"Trammell Et Al vs. Overton." Middle Tennessee Supreme Court, Box 4.

United States. National Park Service. "National Trails System, 'Visit the Trails.'" *National Parks Service.* Accessed November 2, 2014. http://www.nps.gov/nts/nts_trails.html.

"Uses and Abuses of Lynch Law." *American Whig Review of Politics and Literature.* (May 1, 1850): 459-63.

"Vicinity of Fulton, Ark's." *Jeremy Francis Gilmer Papers, #276, 1839-1894.* Southern Historical Collection, Wilson Library, University of North Carolina at Chapel Hill. Accessed September 27, 2014. http://www.lib.unc.edu/mss/inv/g/Gilmer,Jeremy_Francis.html.

Wallace, Ernest, David M. Vigness, and George B. Ward. *Documents of Texas History.* Austin, Tex.: Steck, 1963.

Weather, Pris. "Town of Fulton—1819." *Arkansas Ties.* Accessed September 27, 2014. http://www.arkansasties.com/WhatsNew/2004/11/town-of-fulton-1819.

Weniger, Del. *The Explorer's Texas: The Land and Waters.* Austin: Eakin Press, 1984.

Whitley, Edythe Johns Rucker. *Red River Settlers: Records of the Settlers of Northern Montgomery, Robertson, and Sumner Counties, Tennessee.* Baltimore: Genealogical Pub., 1980.

Williams, Harry Lee. *History of Craighead County, Arkansas.* Greenville S.C.: Southern Historical Press, 1977.

Williams, Jeffrey M. *GIS Aided Archaeological Research of El Camino Real De Los Tejas with Focus on the Landscape and River Crossings along El Camino Carretera.* Master's thesis, Stephen F. Austin State University, 2007.

Williams, Patrick G. *A Whole Country in Commotion: The Louisiana Purchase and the American Southwest.* Fayetteville: University of Arkansas Press, 2005.

Wilson, Shirley. *Sumner County, Tennessee, Bond Book, 1787-1835.* Hendersonville, Tenn.: S. Wilson, 1994.

Winfrey, Dorman H. *Texas Indian Papers, 1825-1843.* Austin: Texas State Library, 1959.

Winters, James Washington. "An Account of the Battle of San Jacinto." *Quarterly of the Texas State Historical Association* 6.2 (1902): 139-44.

Woodruff, Wilford. *Leaves from My Journal . . . Designed for the Instruction and Encouragement of Young Latter-day Saints.* Salt Lake City: Juvenile Instructor Office, 1882.

"Stephen F. Austin." *Texas Vistas: Selections from the Southwestern Historical Quarterly*, edited by Ralph A. Wooster and Robert A. Calvert. Austin: Texas State Historical Association, 1980.

Wright, Muriel H. "Early Navigation and Commerce Along the Arkansas and Red Rivers in Oklahoma." *Chronicles of Oklahoma* 8, no. 1 (1930). Accessed March 1, 2015. http://digital.library.okstate.edu/Chronicles/v008/v008p065.html.

Zuber, W. P. "Thomson's Clandestine Passage Around Nacogdoches." *The Portal to Texas History, The Quarterly of the Texas State Historical Association, Volume 1, July 1897–April 1898.* April 1, 1898; accessed March 1, 2015. http://texashistory.unt.edu/ark:/67531/metapth101009/m1/81/.

INDEX

Adams-Onis Treaty, 96–97, 100, 160
Alamo, 162–64
alcalde, 110, 118–23, 129, 132, 136
Almonte, Juan Nepomuceno (border inspection), 160–61
Alvarado, Luis de Moscoso, 2
Angelina River 3, 30, 61, 111, 121, 125, 128, 132, 173
Anti-Gaming Society, 158
Arkansas Gazette, 8, 109, 114, 169, 174, 179, 191, 199–200, 209, 215
Arkansas Post, 49–51, 68, 80, 85, 92–93, 151
Army, US, 73, 85, 88, 113, 165
Arroyo Hondo, 63, 69, 104
Ashley, Chester, 141–142
Askey, Abner, 132
Askey, Harrison, 91, 187, 207–209, 218
Askey, Lucinda Hill, 91
Askey, Morton (Mote), 40, 44–46, 78–83, 91, 100, 108, 131–33, 136, 144, 187, 207–209 218
Askey, Otho (Otto), 132
Askey, Zachariah, 39, 43, 44–47, 132
Attoyac Bayou, 2, 30, 76, 111, 121
Austin, Moses, 8, 102–104, 111
Austin, Stephen F., 8, 9, 24, 102–107, 110–11, 115–26, 129–30, 134–38, 146–47, 157, 211
Ayish Bayou, 86, 111, 119, 123, 132, 145, 168

Baker, Green, 189, 218
Baker, Nancy Trammell. *See* Trammell, Nancy
Barcroft, Daniel, 175
Barkman, Jacob, 72, 126, 140, 151, 180
Barr, William, 60–61, 74
Batesville (Arkansas), 46, 90, 94, 114, 169
Bayou Pierre, 59, 69
Bean, Peter Ellis, 56, 118, 120–22, 128–30, 146–47
Becca, slave of Nicholas Trammell, 210
bilious pleurisy, 212
blazes, trail markings, 11, 95, 177, 182, 195
Block Brothers, 179, 210
Bloodletting, 212
Boatwright, Thomas, 107
Bodan (Texas), 121
Bois d'Arc Creek, 143
Borderlands, 4–5, 32, 43, 56–59, 64, 67, 69, 73, 77, 84, 98, 112, 135, 158, 166
Borregas Creek, 104
Boston (Texas), 168
Boundaries, 4, 8–12, 18–19, 28–31, 35, 39–43, 52–56, 63–64, 68–73, 88, 96, 97, 103–107, 112–13, 116–19, 123, 129, 139–41, 145–49, 154–56, 160, 167, 172, 183, 192
boundary reserve, 129

bounty warrants, 95, 138
Bowie, James, 141, 161-63
Bowie County (Texas), 18, 100, 172, 188
Boyle, Alexander, 58, 201-203
Bradford, Maj. William, 89, 105
Bradley, John M., 123, 141-42, 215
Brazos River, 106, 125, 185, 186
Bryan, James, 104, 140
Buckner's Creek, 219
Buffalo, 1-2, 26, 35-36, 52, 72-73, 84-85, 89, 92, 100, 105, 117, 162, 174
buffers to immigration, 58, 95, 111-12, 198
Bull's Hill, 128
Bunch, Aden, 115, 120-22
Bunch, Bellona, 120
Bunch, Elizabeth Griggs, 120-22
Bustamante, Anastacio, 104
Byrnside, James, 91, 114-15, 140

Caddo tribes, (see indian tribes)
Cadron (AR), 93, 114
Callier (Collier) family, 137
Camden (Ecore Fabre) (Arkansas), 144, 155, 180, 197, 209, 220
Canebrakes, 16-18, 21, 26-27, 48, 139, 156, 168, 187, 195
Caney Creek, 20, 144
Canoes, 49, 61, 68, 151
Carothers, Hugh, 49
Carrington, Robert, 180-81
Cartwright, Peter, 40-41
Cass County (Texas), 24, 100, 175
chalybeate springs, 25
Charlotte, slave owned by Trammell family, 203
Cherokee tribes, (see indian tribes)
Chicaninny Prairie, 114, 215
Chickasaw Trace, 76
Chicot County (Arkansas), 140, 145, 156, 159, 199-200
Chief Bowles (Duwali), 68, 108-11, 137
Chihuahua Trail, 86, 91, 182

Chisholm, Ignatious, 79
Chisholm, John D., 79-83
Chisholm Trail, 81
Choctaw tribes, (see indian tribes)
chopping axes, 116
Clark's Mill, 180
Clarksville (Texas), 168, 171, 172
Coahuila (Mexico), 110, 124, 129
coal, lignite, 28, 29, 30
Cocke, Watkins, 194
Collier (Callier) family, 137
colonization laws, 8, 33, 103-105, 111, 116, 123-29, 135-36, 146, 166
Colorado River, 103
Congress of the US, 58, 69-71, 85, 94, 107, 138-39, 150-55, 159, 167, 171, 175, 181-83
Conn Creek, 20
Connetoo (John Hill), 68, 79, 194
Contraband, 4, 52, 56, 59-61, 62-69, 84, 104, 126
Conway, James Sevier, 141
Copal, 27
Cordelle, 86
corduroy roads, 151, 178
Crockett, David, 161-63, 167
Cross, Edward 100, 142, 178-79, 197
Crow, Jose Miguel, 67
Custis, Peter, 62
Cypress Bayous, 25-29, 186, 196

Daingerfield (Texas), 172
Dallas (Texas), 109, 183, 210
Dalton Cemetery, 24
Davenport, Samuel, 74
Davidson County (Tennessee), 38, 47
Davis, Daniel, 86, 161, 165, 168, 170
Davis, Zachariah, 126
Dawson, James and Mary, 6, 14, 31, 59
Dayton's Road, 149
de Bastrop, Baron, 103, 111
de Herrera, Simon, 64-66
DeKalb (Texas) 172
Delaware tribes, (see indian tribes)
de Leon, Martin, 119, 146
de Mezieres, Athanase, 2-3

Index

de Moral, Jose Miguel, 58
de Porras, Marin, 66
de Soto, Hernando, 2
de Vaca, Alvar Nunez, 2
Dewees, William, 87–88, 100–106, 116
Dewitt's Colony, 146
Dillard, Nicholas, 100–103, 116
Doak's Stand, Treaty, 100
Dooley's Ferry, (see ferries)
Dubois, Leonard, 124, 125
Dursey, Francis, 88
Durst, John, 118, 125, 136, 173

Eagle Hotel, 201–202, 207
eagle mare, 38
Eaton's Station, 37
Ecore Fabre (Camden), 144, 155–59, 164, 168, 179–81, 186, 190
Eden, Absalom, 115
Edwards, Haden, 126–30, 135–37
Elam's Prairie, 205
El Calle del Norte, 32, 167
El Camino del Caballo, 65
El Camino Real de los Tejas, 3, 5, 32, 55–61, 66–69, 74–75, 84, 97, 104–106, 112–14, 117, 126–31, 140, 145, 158, 161–64, 168–70, 173, 183, 186, 193, 200, 211, 219–20
encarnacion prisoners 193
English, William, 104, 119, 137
Enterprise, steamboat, 147
Epperson's Ferry, (see ferries)
Eskridge, Thomas P., 142
estray bonds, 211

factories, trading posts, 67–68, 80, 93–94, 108, 141
Fannin County (Texas), 172
Faro, 157–58, 201, 204, 207
Fayette County (Texas), 200–203, 207, 212, 219
Featherstonhaugh, George William, 152
ferries, 22, 61, 91, 114–15, 140, 143, 153, 154, 177, 196, 220
    Allen's Ferry, 173
    Byrnside's Ferry, 91, 94, 114–15
    Crow's Ferry, 66–67
    Dooley's Ferry 19, 140, 154–56, 174, 187, 190
    Durst's Ferry, 173
    Epperson's Ferry, 22–23, 154, 171–72, 175, 186, 197
    Fulton ferry crossing, 115, 120, 143–44, 153, 173, 195
    Gaines' Ferry, 113, 196, 215
    Hicks (Hix's) Ferry, 152
    Langford's Ferry, 186, 196
    Lynch's Ferry, 165
    Noland's Ferry, 114, 140
    Pugh's Ferry, 186, 196
    Ragsdale's Ferry, 161
    Ramsdale's Ferry, 28, 175–76, 185–86, 195
    Robbin's Ferry, 91, 173, 193
    Stephenson's Ferry, 24, 107, 175
    Trammel's ferry on the Trinity, 127, 130–35, 140, 193–94, 211
    Walling's Ferry, 176–77, 185
Fields, Richard, 110, 111, 137
filibustering, 55, 77, 96–98, 117, 136, 146, 192
flatboats, 67–68, 76
floating land certificates, 71, 155
floodplains, 17–18, 21–22, 26–27, 87, 101, 128, 139, 143, 172–73, 176, 180
Flores, Gaspar, 129
Forts,
    Fort Jesup, 158, 161
    forts as buffers, 86
    Fort Smith, 83, 89, 93
    Fort Teran, 146–147
    Fort Towson, 113, 148–50, 159–61, 184
    Old Stone Fort, Nacogdoches, 74
Fowler, John, 93, 94, 122
Fowler, Robert B., 141
France, 2, 5, 31, 42–43, 52–53, 72, 76, 101, 117, 179
Frank, slave owned by Trammell, 200–203
Franklin County (Tennessee), 48, 95

fraudulent land claims, 4, 141, 154, 186
Fredonian Rebellion, 135-39, 144, 160
Freebooters, 75, 77, 194
Freeman-Custis expedition, 62, 63, 88, 149
French Lick (Nashville), 37
Fulton (Arkansas), 8-9, 12-18, 24-26, 32-33, 58, 71, 91-96, 101, 107, 113-20, 139, 143-44, 147-48, 152-56, 161-62, 172-79, 182-83, 187, 193-97, 216, 220

Gabriel, slave owned by Trammell, 124, 125, 130
Gaines, James, 112, 118, 136
Galveston (Texas), 97, 220
gambling, 35, 40-41, 91, 122, 127-28, 135, 142, 156-59, 171, 179-82, 186, 199-201, 204, 207, 209, 213
    anti-gambling movement, 158, 171
    baccarat, 157
    billiards, 158
    gamblers, 40-41, 118, 127, 157, 158, 159, 201, 211, 216, 220
    roulette, 127, 156, 157
Georgia settlement, 38
Glass, Anthony, 64
Gold, 2, 51, 124
Gonzales (Texas), 84, 209, 212, 218-19
Gonzales County (Texas), 91, 208, 211-13, 219
Graves, Thomas, 81, 83
Great Raft, 87-88, 147-48, 179
Greenville (Arkansas), 159, 180
Grey, William Fairfax, 162, 167
Grimes, Jemima, (See Trammell, Jemima Grimes)
Guadiana, Jose Maria, 66
guardianship, 40, 47, 204, 206, 208
Gutiérrez-Magee invasion, 73

Hall, Samuel S., 143
Hammond, Mickey, xvi
Hancock, John, 202, 211
Harmony Hill (Texas), 109, 176, 185

Harrison County (Texas), 172, 184, 186, 198
Hawkins, racing challenge, 169-70
Hays, Enoch, 176
Hemphill, Andrew, 125-26, 140
Hempstead County (Arkansas), 96-97, 100, 109, 115, 123, 142-43, 148, 151-59, 168-74, 180-82, 186-91, 210-17
Hendricks Lake, 93
Hill, John (Connetoo), 68, 79
Holloway, Amanda, (See Trammell, Amanda Holloway)
Holston River, 35
horses
    creasing, 54
    crossing rivers, 15, 17, 21-26, 77, 173, 177
    horse racing, 34, 38-41, 91, 121-22, 26, 135, 142, 156-57, 168-71, 179, 182, 187, 195, 201, 207, 213-14
    mules, 3, 15, 57, 64-66, 69, 75, 78, 83-84, 104-105, 110, 116, 121, 187
    mustangs, 5, 18-20, 54, 89
    pack horses, 13-15, 18, 21, 26, 69, 107
    smuggling and trading, 5, 48, 52-57, 65-68, 74, 77-78, 89, 93, 98, 101-106, 109-10, 115, 119, 123-28, 155, 211, 221
    stealing, 4, 11, 18, 24, 40, 53, 60, 69, 77-85, 108, 109, 110, 117, 123, 185, 194
Horton, Alexander, 162, 183, 202, 219
Hoskins, Josiah, 37
Hot Springs (Arkansas), 71, 155, 156, 158, 177, 191
Houston, Sam, 109, 161-65, 182-83
Howard, Benjamin, 78, 81, 83
Hubbard, Thomas, 206
Hughes Springs (Texas), 25, 172
Hunter, John Dunn, 25, 137

Index

Illinois, 91, 107, 210
immigration to Texas, 3, 8, 11–13, 24–25, 32–35, 67–71, 92, 99, 102–107, 111–115, 119, 122–24, 134–35, 139, 143–54, 160, 168, 186, 196, 213, 219–20
improvements to early roads, 4, 12, 16, 58, 76–77, 81, 94–96, 114, 143, 148–54, 159, 167, 178, 183
Independence County (Missouri/Arkansas), 99–100, 116–18, 168, 169
indian tribes,
    accusation of Nicholas Trammell, 77–83;
    capturing mustangs, 55, 60
    Caddo people, 2, 25–27, 72, 85, 93–94, 101, 184
    Cherokee people, 25, 130, 178; relations with Cherokee, 68, 78–83, 86, 94, 108
    Chickasaw people, 185
    Choctaw, Choctaw Line, 113–15, 143, 159
    Choctaw people, 25, 27, 76, 89, 100, 106–108;
    Comanche people, 104–106, 161
    Coushatta, (Alabama & Coushatta) people 220
    death of Nicholas Trammel Sr, 37–39, 47; Delaware people 25, 94, 103, 108, 113–14, 185
    expulsion from Texas, 182–83
    history overlapping with Nicholas Trammell, 108–09
    hopes for land ownership, 109–11
    horses, capturing and stealing, 55, 60, 110
    Houston, Sam, 161–63
    indian agents, 69, 79–81, 88, 137, 186
    Kichai, 53, 130
    Kickapoo, 113
    and and settlements, 31, 37, 48, 58–60, 76, 86, 89, 94–95, 101, 106–107, 112–13, 137, 146, 183, 220
    mounds, 140, 173
    Osage, 80, 83, 85, 89, 93, 107, 108
    relations with other nations, 2–3, 24–25, 52–53, 58, 62, 65–66, 71–72, 80, 86, 89, 95, 109–13, 119, 136–37
    relations with Texas, 162, 163, 183
    Shawnee, 108, 162
    stealing horses, 24, 77–78, 81–83, 109
    Taovaya, 65
    trade, factories, 25, 48, 52–55, 60, 65–68, 75, 83–85, 93–94, 104, 114, 119, 144, 155, 161, 183, 198
    trade, liquor, 82, 144
    trails, 18, 31, 34, 53, 56, 170, 177
    White River, 48, 68, 78, 94.
    See also Fredonian Rebellion
infatuated madmen, 136, 137, 138
inhospitable wilds, 33, 63, 77, 196

Jackson, Andrew, 39, 157
Jackson, John G., 122–23, 141
Jefferson, Thomas, 42, 55, 62–63, 81
Jefferson (Texas), 25, 172, 186, 196
jockey clubs, 157, 168–69
Johnson, Benjamin, 142
Jolly, John, 109
Jones, Anson, 192
Jonesborough (Jonesboro), 8–9, 12, 24, 33, 86, 97–100, 105–107, 113, 116–17, 148–51, 154, 159–61, 168, 171–75, 183, 198

keelboats, 86–87, 93, 147
Kelly, Charles, 92, 138
Kentucky, 4, 8, 24, 34–56, 72, 76, 80, 86, 90–96, 101–102, 108, 116–18, 132, 189, 193, 198–99, 207, 213–14, 220
Kiamichi River, 85, 89, 105–107, 113, 117, 147, 183
kinship, importance to trade, 52, 106, 117

Kiomitia, 183
Kline, David, 49, 50

La Bahia (Goliad), 56, 84, 200-201
La Banita Creek, 32, 167
Lafayette County (Arkansas), 139-43, 148, 154-56, 168, 185, 189-91, 200, 204, 215
Lafferty, John, 48, 70, 79, 80-83, 90-91
Lafferty Creek, 70, 90
Lafitte, Jean, 92-93, 97
La Grange (Texas), 200-202, 211, 219
Lake Comfort, 17, 144
Lamar, Mirabeau B., 182
land offices, 9, 10, 48, 167, 182
Langston, Ragland, 47, 48, 92
Lawrence, Adam, 86, 115
Lawrence County (Missouri), 80, 90-92, 99, 151
Lawsuits, 43-44, 48-49, 92, 125, 141, 199-200, 208, 213-14
Lewis and Clark, 34, 62
liberalization of Mexican immigration laws, 8, 105, 146, 160
Lindsey, Eli, 90
Livingston County (Arkansas), 43-45, 49-51, 80, 118
Logan, Benjamin, 46
Logan County (Arkansas), 38-43, 86, 91, 199
Logansport (Louisiana), 19, 183
Loma del Toro (Bull's Hill), 128
Long, James, 96-98, 117-18, 128
long hunters, 35, 77, 116
Long Prairie, 9, 102, 114, 151, 168
Loring, John, 169-70
Louisiana, 2-5, 12, 42-43, 52-53, 56-58, 62-63, 71-75, 83-84, 87, 92, 96-97, 113, 123, 127, 142, 147, 155-56, 167-68, 171, 180
Lovely, William L., 80-83

Mabbitt, William, 85, 88
Magee, Augustus, 73-74

Magness, Morgan, 99, 155
Magness, Robert, 90, 155
mail routes, 23, 61, 93, 151, 171-72, 178
maintaining old roads, 76, 95, 114, 175-80, 190, 196-97
Mansker's Lick, 36, 130
Mansker's Station, 36
Marion County (Texas), 187
Marriages, 39-40, 90-91, 132, 186-90, 199, 200, 207-208, 211, 215, 218
Marshall, Colonel, 193-95
Marshall (Texas), 172, 176, 220
Martin, Daniel, 176
Martin Creek ,10, 185
Mary, slave owned by Trammell, 189
Mason, Phillip, 37-38
masonic membership, 91
Maulding, Ambrose, 38-39, 47, 95
Maulding, Ennis, 95-96
Maulding, Frances (Fanny), (see Trammell, Frances Maulding)
Maulding, Morton, 36, 47, 92
Maulding, Presley, 164
Maulding, Richard, 39, 47, 95
Maulding, Wesley, 39, 41, 92
Maulding's Station, 36
McGee, Joseph, 64
McKinney, Collin, 140-43
McKinney's Landing, 162
McKinney Bayou 17, 140
McLocklin, Edmond, 119
Meigs, Return, 79-81
Memphis (Tennessee), 150
Mexican independence from Spain, 73, 83, 105-106, 166
Milam, Benjamin, 137, 142, 143, 144, 147, 148, 161
Miles, George W., 201
military, Mexican, 136, 145
military,
    militias, 39, 46, 96, 110, 131-33, 136, 163
    Spanish, 53, 57-66, 92, 98, 108, 117

Index

United States, 4, 12, 46, 69, 73, 83-89, 93-95, 101, 105, 112-13, 138, 150, 158, 161, 193, 196
Military Road (Southwest Trail), 4, 13, 16, 143, 150-53, 174, 177, 178
Miller's Creek, 90
Miller County (Arkansas/Texas), 100, 143, 151, 160, 167, 181
Missions,
  missions, San Jose de los Nazonis, 31
  missions, Spanish, 2, 31, 52, 66
Mississippi, migration from, 68, 89, 196
Mississippi River, 4, 11-12, 42-43, 68-70, 76-78, 86, 108, 140, 150-52, 156, 159, 164, 179, 187, 198-200
Missouri, 4, 8, 34-35, 40, 43-51, 58, 67-73, 77-81, 86, 90-92, 94-95, 99, 103, 105-108, 114, 119, 132, 151-52, 159, 180, 184, 189-90, 193, 198-99, 210
Morrison, Joshua, 120, 140-41, 215-16
Moscoso expedition, 2, 22, 154
Mosquitoes, 21, 26-27
Mound Prairie, 115-16, 206
mounds,
  burial, 2-4, 25-27, 140, 155, 173, 184
  US/Texas boundary, 18-19
Mt. Enterprise (Texas), 30, 220
mules, (see horses)
murder, 40-41, 72, 85, 112, 115, 120-21, 137, 184-86, 190, 198, 207, 214-18
Murrell, John, 215, 216
Musick, Asa, 79, 80, 91
Musick Prairie, 114
mustangs, 3, 5, 8, 18, 53-56, 60-61, 64-66, 73, 77, 84, 100, 105, 108, 117, 121, 139, 155, 168

Nacogdoches (Texas), 2-8, 12-13, 24-33, 52-53, 56-70, 73-77, 86-88, 91-93, 96-149, 154, 158-63, 166-75, 183-87, 190-92, 195-96, 200, 211, 214-15, 220
Naples (Texas), 63
Nashborough (Tennessee), 37, 39, 46
Nashville (Tennessee), 35, 37, 56, 76
Natchez (Mississippi), 53, 55-58, 67, 76, 86, 96, 147
Natchez Trace, 56, 76, 77
Natchitoches (Louisiana), 3, 12, 32, 53-55, 59, 62-64, 67-69, 72-76, 84-85, 88, 91, 94-96, 102-104, 108, 111-14, 143, 147, 151-61, 183, 186, 193, 219
National Historic Trails, 34
Navarro, Jose Antonio, 83, 84
Neches River, 2-3, 61, 128, 147, 162
negroes, 72, 117, 121-26, 189, 203, 205 210
Neutral Ground, 4, 63-64, 67-70, 74, 96, 111, 130-32, 145, 184-86, 198
New Madrid earthquake, 70-72
New Orleans (Louisiana), 42, 51-53, 55, 58-60, 67, 76, 84, 88, 92-94, 147, 170, 179
Nolan, Philip, 55-56, 59-60, 64-65
Noland, Bryan T., 114, 140
Norris, Samuel, 125, 129-33, 136
note of debt, 68, 123, 126, 141-42, 209
Notrebe, Frederic, 51
Nuttall, Thomas, 85-86

O'Quinn, Daniel, 128
Ohio River, 44-45, 76, 86
Oil Trough Bottom, 48, 90
Oklahoma, 8, 88, 113
Ouachita River, 68-70, 156, 179-80, 197
outlaws, 4, 69, 74-75, 83, 89, 105, 109, 137, 158, 171, 190, 198-99, 215
over-hunting, 94, 109
overseer, road maintenance, 95, 144, 177, 180, 197-98, 215
Overton, John, 47-51, 92

Panola County (Texas), 9, 176
Parmer, Isom, 144-45, 187, 211
Parmer, Martin, 132, 136-37, 142-45, 155, 187
Passports, 55, 66, 72, 147
Pecan Point (Texas), 8-9, 12, 24-25, 31-33, 71-73, 84-90, 94-113, 116-19, 130-33, 137, 143, 146, 148-49, 154-56, 160-62, 165-70, 175, 182, 198, 211
Pennington, Isaac, 114, 115
picarros on the Sabine, 119, 130
Pinkerton, Joan Strong, xvi
plank roads (turnpikes), 151-52, 178, 197, 220
plantations, 18, 86, 131, 140, 156, 181, 187-88, 204-205
Poer, Martin, 164
Poke Bayou, 80, 114
Polk, Benjamin, 193
Pope, John, 150
population numbers, 32, 146, 166-68, 192
ports, 32, 96-97, 144, 156, 172, 178, 187
post offices, 140, 151, 171, 196, 207
Potter, Robert, 186
prairies, 8-9, 21, 42, 48, 61, 66, 89, 95, 102, 109, 114-16, 121, 139-41, 151, 154-58, 162, 168, 173-74, 184-87, 205, 215, 220
    Bloody Prairie, 184-85
    Blossom Prairie, 168
    Bodark Prairie, 115
    Chicaninny Prairie, 114, 215
    Fisher's Prairie, 114
    Hog Prairie, 95
    Horse Prairie 89
    Jonesborough Prairie, 173
    Long Prairie, 8-9, 102, 114
    Lost Prairie, 109, 114, 139-41, 151, 154-56, 158, 162, 174, 187, 205
    Mound Prairie, 115-16
    Mustang Prairie, 61, 66, 121

Red River, 5, 8, 18, 20, 24, 55, 65, 77, 113, 168, 195
    Sulphur Fork, 18-21, 148-49, 153, 162
pre-emption land certificates, 46-47, 138, 154-55
Prescott (Arkansas), 220
Prieto, Pedro Lopez, 65
probate court, 204-206, 211
Procela, Luis, 129
Pryor, William, 125
public land grants, 46, 71, 79, 85, 95, 154, 155, 159, 190
Puelles, Jose Maria, 7
Pulaski County (Arkansas), 141, 172
puncheons, 20, 151

Quinn, Daniel, 125
Quirk, Henry, 64

racetracks, 121, 157-58, 168-71, 201
rafts, for crossing rivers, 15, 22
Ramsdale, Francis, 175
Ramsdale's Ferry, (see ferries)
Red River, Great Bend, 3-4, 8, 14-16, 63, 88, 96-98, 116, 173
Regulator-Moderator War, 186
Religion, 23, 179-80, 193
    Ministers, 42, 89-90, 196
    Methodist, 40-41, 89-91, 168, 190
    Presbyterian, 42
republic of Texas, 8-9, 18-19, 23, 30, 54, 96, 137, 164, 167-68, 171-77, 182, 183-88, 192
Revolutionary War, 35, 46
Revolution, 96, 143, 161-64, 173, 183, 187, 192
river fords 14-15, 22, 27-29, 102, 107, 154, 175
roads,
    conditions, 4, 9, 11, 16, 58, 61, 69, 76, 95-96, 107, 114-15, 145, 148-53, 159, 171, 176, 179-80, 213, 219-20

plank roads (turnpikes), 151-52, 178, 197, 220
road-building, 4, 11, 58, 76, 81, 95-96, 143, 148-53, 159, 171, 177-81, 190, 196-97, 213, 219-20
smugglers, secret roads, 56-57, 60-67, 93, 128
Robbins, Nathaniel, 91, 173, 193
Roberts, David, 115
Rocky Mound, 220
Rodgers, William C., 49
Rogue's Harbor, 40-41
Rosborough Springs, 184
Rose, William Pinckney, 186
Runaway Scrape, 164
Rusk County (Texas), 9-10, 31, 172, 177
ruts, across landscape, 3-4, 9, 12, 20, 23-26, 29, 152, 178, 220

Sabine River, 3-5, 19, 28-30, 53-54, 57-59, 62-63, 66-69, 73-75, 86, 93-98, 101-105, 110-13, 117-19, 132-33, 136-37, 145-47, 154, 158, 161, 167, 175-77, 183-86, 215, 220
Sadberry, Marvin T., 190
Sadberry, Mary, (see Trammell, Mary Sadberry)
Salcedo, Nemesio, 57, 66
salt licks, salines, 32, 35-36, 40, 47-49, 52, 61, 76, 86, 125, 130
Samuels, Robert (Uncle Bob), 215
San Antonio, 3, 8, 57-60, 65, 69, 73, 83-84, 97, 102-106, 110, 127, 183, 193, 200
San Augustine (Texas), 123, 127, 145, 167-72, 185, 219
San Jose de los Nazonis, 31
Sanchez, Luis, 131
Sanchez, Mariano, 31, 131
San Felipe de Austin, 124
San Jacinto (Texas), 111, 164, 165
Santa Anna, 164, 192
Santa Fe (New Mexico), 34, 62, 91-93, 182, 192

Sartuche, Ignacio, 128-29, 133
Sawyer, Edmond, 201, 203
scarifying and cupping, 212
Seguin, Jose Erasmo, 103, 106
Seguín, Juan, 118
Sevier, John, 152
Shaw's Creek, 115
Shawneetown, 158
Shelbyville (Texas), 172
Sheriff, 39-41, 45, 51, 92, 132, 138, 141, 145, 154, 202, 215
Shipman, Daniel, 107
shoals, used for crossing rivers, 14, 22, 67, 73, 84, 148, 154, 173-75
Shreveport (Louisiana), 186
Shropshire, Benjamin, 219
Sibley, John, 69-73, 88, 105
slaves, 4, 38, 45, 112, 119, 122-30, 133-37, 140, 146, 156, 163, 166, 181, 187-92, 196, 199-219
sloughs, 20-21, 26, 197
Smith, Jabez, 209-10
Smith, William, 79-80
smuggling, 4-5, 16-20, 32-35, 43, 52-69, 73-78, 83-84, 92, 98-99, 104, 105-106, 110, 116-17, 131-34, 146, 155-57, 166, 180, 186, 191-94, 197, 207, 220-21
Sophia, slave owned by Trammell, 125, 126
Southwest Trail, 4, 11, 58-59, 70, 75, 95, 107, 114, 138, 143, 150-53, 156-59, 162-64, 174-79, 183, 196
Spain, 2-8, 24-26, 31-34, 43, 51-77, 83-85, 88, 92-113, 117, 124-27, 133, 149, 160, 166, 175, 198, 215, 220
Spanish Bluff, 88, 107, 113, 149, 198
Spring Hill (Arkansas), 180, 181, 190
Springs, 1-2, 11, 18, 25, 29-30, 35-36, 47, 162, 178, 184-85, 207
Sprowl, John, 123
St. Denis, Louis Juchereau de, 53, 76
St. Louis (Missouri), 4, 58, 70, 114, 151, 156, 182-83
Steadham's racetrack, 121

steamboats, 147, 177–79, 220
Steele, John M., 90
Sterne, Adolphus, 172
Stevenson, William, 89–90, 114
streams, 14, 20, 30–32, 63, 88, 107, 144, 150–52, 162, 190
stumps in roadways, 12, 25, 87, 95, 116, 152, 171, 180, 196
Styles, William, 89, 105
Sugar Hill plantation, 18
Sulphur Fork of the Red River, 18, 21–24, 93, 107–108, 149, 154, 167, 175, 186, 197
supreme courts, 92, 96, 201, 203, 208, 218
surveying land, 9–10, 18, 31, 45–48, 62, 92, 127129, 133, 140, 151, 160, 175, 183–84, 187, 213
swamps, 11–12, 20–21, 27, 61, 88, 150, 195
swimming rivers, 15, 61, 173

Tate, James, 127, 140
Tatum (Texas), 176
Tavern,s 76, 180, 193, 216
    Hays', 176
    Parmer's, 144, 155
    tavern permits, 115, 144
    Trammell's, 155–59, 164, 179–81, 186–90, 195, 197–203, 210, 212–14, 217–20
    Vaughn's, 159
Tennesseans, 77, 89–90, 99–101, 147, 154, 161–62
Tennessee, 4, 8, 24, 34–51, 56, 68, 72, 76, 80–81, 86, 90–92, 95, 99, 107, 116–19, 132, 147, 150, 161–65, 193, 198, 213
Terán, Manuel de Mier y, 134, 145–47, 161
Terre Rouge, 156
Texarkana (Texas), 4, 18
Texas,
    boundary with United States, 19, 174

immigration, colonization, 8, 12, 35, 68, 100–108, 114–18, 126, 166–68, 174, 196, 212
independence, republic, 23, 84, 123, 132, 136–39, 143, 158–68, 171–73, 183–88, 192, 198
Mexican, 8, 12, 108, 112, 116–17, 120, 123–30
Spanish, 2, 4–5, 8, 43, 51–53, 55–58, 62–63, 69–77, 83–85, 92–98, 101–103
Thomas Creek, 23
Thurmond, Celia, 199
Thurmond, Julia Ann, 188, 199–200, 209
Thurmond, Rebecca, 199
Thurmond, Richard, 199
Thurmond, William, 199
Tiller, Nancy, 14, 31, 59
Touch, Sally 188,
trace, definition, 5
trading posts, 48–51, 68, 80, 84–85, 92–94, 103, 114, 117, 151
Trammell, Amanda Holloway, 207, 208
Trammell, Eliza Lindsey, 200
Trammell, Frances (Fanny) Maulding, 35–40, 46–47
Trammell, Henry, 140, 156, 188, 191, 199–200, 204–209, 218–19
Trammell, Jarrett, 206, 208
Trammell, Jemima Grimes, 38
Trammell, Leona Ward, 200, 203
Trammell, Mary Sadberry, 190, 210–13, 218–19
Trammell, Nancy Baker, 189, 218
Trammell, Nathaniel, 127, 128, 140, 141, 144, 156, 187, 200–204, 207–11, 218–19
Trammell, Nicholas,
    expulsion from Texas, 131–39
    horse racing, gambling, taverns, 41, 156–59, 168–73, 179–81, 186, 197, 202–207, 218–20
    land ownership, 44–48, 70–73, 80, 90–92, 116–20, 127–31,

Index

138–40, 155–59, 187–90, 199, 211
   legal involvement, 38, 44–45, 48–50, 92, 100, 118–26, 130–32, 139–42, 188, 200–10
   myths about Nicholas Trammell, 189–90, 194, 214–16, 218
   naming of Trammel's Trace, 9, 34–35, 79, 101, 116–17, 147
   obituary, false for Trammell, 207
   religion, 90–91
   slave ownership, 123, 125–26, 187–90, 200–10, 219
Trammell Sr, Nicholas (father), 35–39, 46–47
Trammell, Nicholas (grandson), 204–208
Trammell, Nicholas (son), 211, 218–19
Trammell, Phillip (grandfather), 38
Trammell, Phillip (son), 121–22, 140, 156, 187, 190–91, 200–209, 219
Trammell, Phillip (uncle), 40, 86
Trammell, Robert, 187
Trammell, Sarah (daughter), 211, 218
Trammell, Sarah (wife), 50, 130, 187–90, 218
Treaties, 63, 81, 96–100, 106, 113, 160–62, 183
Trespalacios, Jose Felix, 110
Trimble, James, 45
Trimble, William, 79–80, 83, 90, 142
Trinidad de Salcedo, 61–66, 128
Trinity River, 32, 57–66, 84, 91, 109, 121, 126–35, 140–47, 165, 173, 183, 189, 193–94, 211, 220
turnpikes, early toll roads, 178–79, 197

Ugarte, Jose Joaquin 59, 60
United States, 3–5, 8, 11–19, 24, 32, 39, 42–43, 53–116, 128–34, 137–72, 178–85, 190–93, 197, 205

Vaughn, Polly, 156–59
Viana, Francisco, 63
Villon, Francois, 124–25

Virginia, 35–36, 39, 42, 102, 181

wagons, 3, 12–16, 18–21, 24–32, 42, 69, 77, 80, 89, 95, 116–18, 143–45, 149–53, 159, 164, 173, 177–82, 192–95, 212–13, 218
Walling, John, 176, 177
Ward, Leona, (*see* Trammell, Leona Ward)
War of 1812, 84
Warrants, 46–47, 95, 138, 145
Washington (Arkansas), 12–13, 114–15, 143–44, 148, 150–52, 155–59, 162–64, 168–70, 174, 177–81, 186, 193–96, 220
Washington (DC), 96
Wavell, Arthur, 143
Wetmore, Alex & George, 85
Whatley Creek, 23
whiskey, 80–81, 94, 109, 180
White, Solomon, 38
White's Creek, 37
White River, 45–49, 70, 73, 78–81, 86, 90–94, 99, 100, 108, 114–18, 132, 138–39, 151, 155, 169, 214
wife, wonderful (*see* Hammond, Mickey)
Wilkinson, James, 63–64
Williams, Sam, 215–17
Williams Settlement, 30–31
Wright, Claiborne, 85–86, 90, 183

Y'Barbo, Jose Ignacio, 59
Yoakum (Yokum) family, 137
York, Ellison, 122
York, John, 118
Young, John, 45, 68
Young, Peter, 120–21

Other titles in the Red River Books series:

Beyond Redemption:
Texas Democrats after Reconstruction
by Patrick G. Williams

Planting the Union Flag in Texas:
The Campaigns of Major General Nathaniel P. Banks in the West
by Stephen A. Dupree

The Great Southwest Railroad Strike and Free Labor
by Theresa A. Case

Red River Bridge War:
A Texas-Oklahoma Border Battle
by Rusty Williams

CPSIA information can be obtained
at www.ICGtesting.com
Printed in the USA
LVHW040943310122
709513LV00003B/4